FERMI REMEMBERED

FERMI REMEMBERED

Edited by

James W. Cronin

The University of Chicago Press

Chicago & London

James W. Cronin is University Professor of Physics and Astronomy Emeritus at the University of Chicago. He was a graduate student at the University of Chicago from 1951 to 1955.

The University of Chicago Press, Chicago 60637
The University of Chicago Press, Ltd., London
© 2004 by The University of Chicago
All rights reserved. Published 2004
Printed in the United States of America
13 12 11 10 09 08 07 06 05 04   1 2 3 4 5

ISBN: 0-226-12111-9 (cloth)

Library of Congress Cataloging-in-Publication Data
Fermi remembered / edited by James W. Cronin.
    p. cm.
  Includes bibliographical references.
  ISBN 0-226-12111-9 (cloth : alk. paper)
  1. Fermi, Enrico, 1901–1954—Archives.   2. Nuclear physics.   I. Fermi, Enrico,
1901–1954.   II. Cronin, James W., 1931–
  QC16.F46F49   2004
  539.7'092—dc22
                                                                                2003020524

♾ The paper used in this publication meets the minimum requirements of the
American National Standard for Information Sciences—Permanence of Paper for
Printed Library Materials, ANSI Z39.48-1992.

Contents

Preface

I arrived at the University of Chicago for graduate study in 1951, and was very fortunate to attend many classes that were taught by Enrico Fermi. Fermi was indeed a great teacher, but he also had a way of ensuring that one understood the lecture. At the end of each lecture, he orally assigned a problem that had to be worked and handed in at the next class. These were then very carefully graded by an advanced graduate student. Real grades were assigned to those who took his courses. When it became clear that we at Chicago had to have a significant recognition of Enrico Fermi's hundredth anniversary, I eagerly took the responsibility to organize it. Fermi's presence during my graduate years was important not only for direct contact with him, but also for the exposure to all the extraordinary physicists that were attracted to Chicago by his presence. During the years 1945–54 there was certainly no better physics department in the world. This environment, whether by osmosis or by other mysterious processes, instilled in me and many of my student colleagues a passion for physics which has lasted a lifetime.

September 29, 2001, was the hundredth anniversary of Enrico Fermi's birth. On this exact date we organized a symposium at the University of Chicago to remember him. Fermi's contributions to science and technology are vast. For this symposium we decided to emphasize the years 1945–54 in Fermi's life, when he was professor of physics at the University of Chicago. With this theme, we invited Fermi's faculty colleagues, Fermi's students, and students who were studying at the Institute for Nuclear Studies during this period. The response to this event was enthusiastic, and a number of Fermi's colleagues gave talks remembering him. Many other colleagues who could not attend submitted reminiscences.

The organization of the symposium was ably arranged by Caryn Myers and Patricia Plitt. University of Chicago president Don Randel provided the resources for the symposium, gave the welcome address, and sat through most of the symposium, thoroughly enjoying it. It was a pleasure to work with Laura Grillo on the financial aspects of the symposium.

While the focus of the event was on Fermi's Chicago days, inevitably aspects of the first controlled self-sustaining nuclear reaction at Chicago, and the years at Los Alamos during World War II, are mentioned in many of the contributed articles. This was because so many of the students and faculty colleagues of Fermi came under his influence during the war period.

In preparation for the meeting, I spent many hours in the archives going through Fermi's papers. In so doing I found many letters, notes, and outlines of speeches which beautifully revealed Fermi's character. I assembled this material in multiple series of notebooks that were available for browsing during the social hours of the symposium. I originally thought of just putting together a volume of reminiscences, but I realized that, by adding several chapters containing material from the archives, a more interesting volume would result. While the emphasis of the book is devoted to the Chicago years, I chose to include a short biography of Fermi written by Emilio Segrè, reprinted from *Fermi's Collected Papers*. This is complemented by an essay on Fermi's impact on physics by Frank Wilczek, a distinguished young physicist who was born three years before Fermi's death. Then follow three chapters containing material from the archives. These are, respectively, letters and documents concerning the development of nuclear energy, letters sent and received by Fermi, and documents concerned with his research and teaching.

The next three chapters contain articles written by Fermi's faculty and research colleagues, Fermi's students, and students whose thesis sponsors were other than Fermi, but who were in that magic environment at the Institute for Nuclear Studies during the Fermi years. Finally, I have concluded the book with an essay which compares Fermi's predictions for the future of particle physics with what has actually happened. Twenty-seven individuals have contributed to the volume; brief biographies of each are given at the end.

There are many individuals who must be thanked for their help in the preparation of this volume. From the Joseph Regenstein Library Department of Special Collections, which houses the Fermi archives, I acknowledge the enthusiastic help of curator Alice Schreyer, archivist Daniel Meyer, exhibitions manager Valarie Brocato, and Eileen Ielmini, who had just reorganized the Fermi archives. Judith Dartt was responsible for scanning and placing on CDs all the precious archival material.

While I was searching the archives I met Giulio Maltese, who was on the same mission from Italy, and who pointed me to some gems in the archives that I might have missed. I would also like to thank librarian Kathleen Zar and physics bibliographer Fritz Whitcomb, of the John Crerar Library, for their assistance.

Participation in the preparation of the manuscript far beyond the call of duty has been provided by my secretary, Sandy Heinz. I want to thank all the speakers, and those Fermi colleagues who could not attend the symposium, for the prompt submission of their manuscripts. I thank the myriad of individuals and institutions that have provided the permission to publish photographs and letters. The responsibility for long delay in the preparation of this book cannot be placed with any individual.

Everything takes time. My editors at the University of Chicago Press, Christie Henry and Jennifer Howard, have been easy to work with. My copy editor, Erik Carlson, gave careful attention to details I never thought existed.

Chapter One

Biographical Introduction

This first chapter presents a brief biography of Enrico Fermi which was written by Emilio Segrè for *Fermi's Collected Papers* (1962). While this volume is concerned with Fermi's life in the United States with much emphasis on his years at the University of Chicago, more than half of his scientific life was spent in Italy. To gain an appreciation of Fermi's total impact on physics, we have chosen to reprint the biography. The references given at the end of this chapter conform to the numbering in Fermi's collected papers.

❖ ❖ ❖

Emilio Segre
BIOGRAPHICAL INTRODUCTION

Enrico Fermi was born September 29, 1901, in Rome at Via Gaeta N° 17, the third child of Alberto Fermi and Ida De Gattis. His family originally came from the North Italian town of Piacenza. He grew up in Rome, where he attended grammar school and later the Ginnasio Liceo Umberto I. He then studied at the University of Pisa as a

First published in *Enrico Fermi: Collected Papers*, volume 1 (University of Chicago Press, 1962), pp. xvii–xlii.

Fellow of the Scuola Normale Superiore, and in 1922 he received his doctorate in physics. In 1923, with a Fellowship of the Italian government, he went to Göttingen to work under Max Born; and in 1924 he moved to Leyden to work under P. Ehrenfest. He returned to Rome the same year and in 1925 became "Libero Docente". He then held a lectureship at the University of Florence, and finally, in 1926, he was appointed Professor of Theoretical Physics at the University of Rome, where he remained until 1938.

In 1938 the Fascist racial laws, which affected his wife Laura Capon and her relatives and moreover deeply offended his sense of fairness, induced him to emigrate to the United States of America. Here, he was at first Professor of Physics at Columbia University in New York; then, during the Second World War he devoted all his activity to the development of nuclear energy on a large scale at Chicago, Illinois and at Los Alamos, New Mexico. At the end of the war, in 1946 he moved to the new Institute for Nuclear Studies at the University of Chicago, where he remained until his early and unexpected death on November 29, 1954.

These are the bare, external facts of Fermi's life. As is often the case with a man of science, they do not seem especially dramatic, although in his case the development of nuclear energy, more than emigration to America, certainly represented an event comparable in adventure to the discovery of a continent for an explorer.

The lamented circumstance of his early death has as a consequence that many of his contemporaries, collaborators, and friends remember him well. In recounting their experiences with him they can try to reconstruct a living image of the man. This is highly desirable because many scientists have the natural desire to know personally, as it were, the major worthies of their craft. What physicist has not had the curiosity to learn what kind of a man Galileo, Newton or Maxwell was? How did he work? How did his ideas originate? These questions are not easily answered by the ordinary biographies, generally written by authors with little personal and scientific contact with their subject. In Fermi's case, however, the fact that he collaborated extensively with other scientists, who were often his own pupils, makes it easier to reconstruct his scientific portrait. For this reason, while the single papers here collected speak for themselves scientifically and give the finished product of his work, we have tried as often as possible to provide for each of them an introduction written by whoever is in a position to be well informed of the circumstances which led to that specific investigation. While these introductions can add but little to the scientific interest of the papers, we hope that they may contribute to a reconstruction of the picture of the scientist as we knew him.

If we examine Fermi's works as a whole, we find them still fresh and interesting, often written with rare pedagogical skill. In a certain sense, the papers make easy reading, so that an able student can profit greatly from their study, even as a beginner; at the same time, they often have such deep implications that they have inspired very difficult and recondite investigations. The typical style of Fermi's writings is a close reflection of his personal and intellectual history. We see it evolve as time goes on from the early papers written at Pisa, in almost complete isolation, on a variety of disconnected problems, to the late papers in Chicago when he was a mature physicist and a great leader in the field, investigating theoretically and experimentally, with numerous co-workers, a whole new chapter of physics.

In the following pages I shall try to render Fermi's image as it appears to me.

Fermi's interest in mathematics and physics manifested itself very early. He told me that when he was about ten years old he had seriously struggled to try to understand why a circle was represented by the equation $x^2 + y^2 = R^2$, and had finally succeeded only after great efforts. This episode gives us a measure of his development at that age.

A little later he must have made considerable progress because while still an adolescent he read, in Latin, a book entitled "Elementa Physicae Mathematicae" by "Andrea Caraffa e Societate Jesu, in Collegio Romano". This is an ample treatise of mathematical physics published by a Jesuit in 1840. Its two volumes comprise approximately 900 pages and they cover mechanics, optics and astronomy, using the resources of analysis. Fermi acquired the book in a used-book stall and must have studied it quite diligently because the book has numerous marginal notes and developments of the calculations in his handwriting.

We know that what little outside help Fermi received came mostly from Ingegnere A. Amidei who was among the first to notice the boy's extraordinary ability. In this connection we have the invaluable first-hand testimonial of Ing. Amidei as set forth in a letter by him to E. Segrè. He states that sentences quoted verbatim were noted by him in writing at the time of his conversations with Fermi. We quote here Ing. Amidei's letter:

"Antignano (Livorno).
"November 25th, 1958.
"Having had the opportunity of guiding and counseling Enrico Fermi in his studies during his youth from his 13th to the end of his 17th year, I deem it appropriate to hereby summarize his course of study during the

above-mentioned period, which was from the autumn of 1914 to the autumn of 1918.

"In 1914 I was Principal Inspector of the Ministry of Railways, and my colleague was Chief Inspector Alberto Fermi. When we left the office we walked together part of the way home, almost always accompanied by the lad Enrico Fermi (my colleague's son), who was in the habit of meeting his father in front of the office. The lad, having learned that I was an avid student of mathematics and physics, took the opportunity of questioning me. He was 13 and I was 37.

"I remember very clearly that the first question he asked me was: 'Is it true that there is a branch of geometry in which important geometric properties are found without making use of the notion of measure?' I replied that this was very true, and that such geometry was known as 'Projective Geometry'. Then Enrico added, 'But how can such properties be used in practice, for example by surveyors or engineers?' I found this question thoroughly justified, and after having tried to explain some properties that had very useful applications, I told him that the next day I would bring him—as I did—a book on projective geometry by Professor Theodor Reye [1] that included an introduction which, in a masterful, artistic style succeeds, by itself, in explaining the usefulness of the results of this science.

"After a few days Enrico told me that besides the introduction he had already read the first three lessons and that as soon as he finished the book he would return it to me. After about two months he brought it back, and to my inquiry whether he had encountered any difficulties, replied 'none', and added that he had also demonstrated all the theorems and quickly solved all the problems at the end of the book (there are more than two hundred).

"I was very surprised and since I remembered that I had found certain problems quite difficult and given up trying to solve them since they would have taken too much time, I wanted to verify that Enrico had also solved those. He gave me the evidence.

"Thus it was certainly true that the boy, during the little free time that was left to him after he had fulfilled the requirements of his high school studies, had learned projective geometry perfectly and had quickly solved many advanced problems without encountering any difficulties.

"I became convinced that Enrico was truly a prodigy, at least with respect to geometry. I expressed this opinion to Enrico's father and his reply

(1) T. Reye, *Geometria di Posizione*. Tr. A. Faifofer, Venezia 1884.

was, yes, at school his son was a good student, but none of his professors had realized that the boy was a prodigy.

"I then learned that Enrico studied mathematics and physics in second-hand books that he bought at Campo dei Fiori, hoping to find one treatise that would scientifically explain the motion of tops and gyroscopes, but he could never find an explanation, and so, mulling the problem over and over again in his mind, he succeeded in reaching an explanation of the various characteristics of the mysterious movements by himself. Then I suggested to him that to obtain a rigorous explanation, it was necessary to master a science known as 'Theoretical mechanics'; but in order to learn it he would have to study trigonometry, algebra, analytical geometry and calculus, and I advised him not to try the problems of tops and gyroscopes, since he would be able to solve them easily once he had mastered the field that I had outlined. Enrico was convinced of the soundness of my advice and I supplied him with the books that I thought were most suitable for giving him clear ideas and a solid mathematical base.

"The books which I loaned him and the date of the loan are as follows:

"In 1914, for trigonometry, the treatise on plane and spherical trigonometry by Serret.

"In 1915, for algebra, the course on algebraic analysis by Ernesto Cesàro and, for analytic geometry, notes from lectures by L. Bianchi at the University of Pisa.

"In 1916, for calculus, the lectures by Ulisse Dini at the University of Pisa.

"In 1917, for theoretical mechanics, the 'Traité de mécanique' by S. D. Poisson.

"I also deemed it appropriate for him to study the 'Ausdehnungslehre' by H. Grassmann which has an introduction on the operations of deductive logic by Giuseppe Peano. These books were loaned to him in 1918.

"I thought it appropriate because it was my opinion that the Ausdehnungslehre (similar to vector analysis) is the most suitable tool for the study of different branches of geometry and theoretical mechanics. . . .

"Enrico found vector analysis very interesting, useful and not difficult. From September, 1917, to July, 1918, he also studied certain aspects of engineering in books that I lent him.

"In July, 1918, Enrico received his diploma from the Liceo (skipping the third year) and thus the question arose whether he should enroll at the University in Rome. Enrico and I had some long discussions on this subject.

"First of all I asked him whether he preferred to dedicate himself to mathematics or to physics. I remember his reply and I transcribe it here literally: 'I studied mathematics with passion because I considered it necessary for the study of physics, *to which I want to dedicate myself exclusively'*. Then I asked him if his knowledge of physics was as vast and profound as his knowledge of mathematics and he replied: 'It is much wider and, I think, equally profound, because I've read all the best known books of physics' (*). I had already ascertained that when he read a book, even once, he knew it perfectly and didn't forget it. For instance, I remember that when he returned the calculus book by Dini, I told him that he could keep it for another year or so in case he needed to refer to it again. I received this surprising reply: 'Thank you, but that won't be necessary because I'm certain to remember it. As a matter of fact, after a few years I'll see the concepts in it even more clearly, and if I need a formula, I'll know how to derive it easily enough'.

"In fact, Enrico, together with a marvelous aptitude for the sciences, possessed an exceptional memory.

"I then considered that the proper moment had arrived to present a project that I had already considered in his behalf for a year, that is from the time when I advised him—and he immediately followed my suggestion—to study the German language, since I foresaw that it would be very useful for reading scientific publications printed in German without having to wait until they were translated into French or Italian. My plan was this: Enrico ought to enroll not at the University of Rome, but at the one in Pisa. First, however, he would have to win a competition to be admitted to the Scuola Normale at Pisa and attend (besides the courses in the School) the University of Pisa.

"Enrico recognized at once the soundness of my plan and decided to follow it, even though he knew that his parents would be opposed to it. Then I immediately went to Pisa to obtain the necessary information and the program for the competition to the Scuola Normale Superiore and immediately returned to Rome to examine it minutely with Enrico. I ascertained that he knew the subjects in the group of mathematics and physics perfectly, and I expressed my conviction that he would not only win the competition, but would also be the first among the applicants—as in fact he was.

"Enrico's parents did not approve of my plan, for understandable and

---

(*) Persico says that one of them was the French translation of the large Treatise by Chwolson. During the Summer of 1918, Fermi went almost every day to the library of the "Instituto Centrale di Metereologia e Geodinamica" in order to study this book. Permission had been granted by the Director Prof. F. Eredia who had been Fermi's physics teacher at the Liceo Umberto I.

human reasons. They said: 'We lost Giulio (Enrico's older brother, who died after a short illness in 1915) and now we are to allow Enrico to leave us to study at Pisa for four years while there is an excellent university here in Rome. Is this right?'

"I used the necessary tact to persuade them, a little at a time, that their sacrifice would open a brilliant career for their son and they finally agreed that my program should be carried out. Thus, as Enrico's wife wrote in her book 'Atoms in the Family', 'at the end, the two allies—Fermi and Amidei—carried the day".

Thus far goes the Amidei letter.

Persico tells of experiments undertaken together with Fermi at about the same time as that covered by the Amidei letter [2]. They are noteworthy especially for the choice of problems, such as to determine with high precision the density of Rome's drinking water (Acqua Marcia). This is a rather different problem from the more common one of a boy wanting to build some sort of an electric motor or apparatus: today this would be most probably a radio; at the time we are considering, it would have been a wireless telegraph. This does not preclude the fact that, at the time, Fermi was an expert builder of electric motors, but obviously he was also interested in more sophisticated problems such as the investigation of the behaviour of gyroscopes.

The point reached by his scientific development at the age of 17, when he finished his secondary schools, is clearly demonstrated by the entrance examination for the "Scuola Normale Superiore" at Pisa. The theme was "Distinctive properties of sounds" and is dated November 14, 1918. Certainly the examiners expected an essay at a high school level as one reasonably would. Instead, after a few introductory sentences, we find in his examination sheet the partial differential equation of a vibrating reed and its solution with the help of a Fourier series. The examiner, Professor G. Pittarelli, a very kindhearted man and a good mathematician in his own right, must certainly have been stunned by the little essay. Fermi himself told me that Pittarelli, after having read the examination paper, called him to ascertain whether the candidate really understood what he had written. After questioning him, Pittarelli said that during his long teaching career he had never met anybody like Fermi and that certainly the boy had extraordinary talents. Fermi remembered these occurrences many years later with obvious pleasure and deep gratitude towards Pittarelli.

In his other studies it is clear that at the Ginnasio and Liceo he pro-

---

(2) E. PERSICO, "Scientia", vol. 90, pag. 36 (1955).

gressed brilliantly and without effort. He was the sort of model student who succeeds in everything. His professor of Italian was Giovanni Federzoni, and Fermi who had an exceptional memory, knew long excerpts from the "Divine Comedy" and other Italian poems by heart and for the rest of his life was able to quote them on appropriate occasions. "Orlando Furioso" by Ariosto apparently was one of his favorite readings even before he had to read it at school; already when he was about 12 years old he could recite entire cantos by heart. For him one reading was sufficient to commit a section to memory. He must have been bothered by Greek, which was compulsory in Italian schools. At Los Alamos I once complained that in a sort of nightmare I had dreamed of a Greek final examination at the Liceo: Fermi confessed that he had been subject to the same nightmare.

His literary tastes were very simple; his own writings, including popular lectures, are not noteworthy for elegant literary style, especially in his early years. He had little sensitivity to literary form; to him the content was the only thing that mattered. On the other hand he was extremely careful, almost pedantic, as far as scientific precision was concerned. This care for precision increased steadily with time and one notices a remarkable difference in style between the early papers and those of his more mature years.

With regard to his knowledge of languages, he learned German as a boy, as Ing. Amidei mentions. The complexities of German grammar fascinated him. I think in his youth he could sit down and write an article in German without any mistakes, and he spoke it fluently; he knew French as many Italians do, using it easily but not always correctly. More than any other language except Italian he used English. He learned everything that one can by application and study, but the muscles of his mouth never became accustomed to English sounds, and he always retained a strong Italian accent, which occasionally irritated him. In America, Fermi devoted more effort to his English pronunciation than most immigrants are wont to do, but the result always remained imperfect.

Fermi registered at the University of Pisa as a Fellow of the Scuola Normale Superiore in the fall of 1918. This school was founded by Napoleon in 1813 as a branch of the École Normale Supérieure of Paris. Its original purpose was the preparation of high-school teachers and the promotion of higher studies and research. Its pupils attend the University of Pisa, but in addition have special courses, mostly of the character of seminars. The pupils and professors live in the school as in a British college. Admission is by competition only and there are no fees of any kind. Several of the most distinguished literary and scientific figures of modern Italy studied there, and the roster of its alumni adds to the great prestige of the institution.

Fermi chose mathematics as his major subject but soon changed to physics. He conducted his studies at Pisa very independently. His grasp of physics was far above the local level of teaching and from his correspondence with Persico [3] we have some detailed information on how he spent his time. In February of 1919, as a freshman, he writes: "Since I have almost nothing to do for my class work and I have many books available, I try to enlarge my knowledge of mathematical physics and I shall try to do the same for pure mathematics". In the same correspondence he says that he has read Poincaré's "Théorie des tourbillons" and Appell's "Mécanique rationnelle" devoting himself especially to the methods of analytical mechanics. He also started studying Nernst's "Theoretische Chemie" and the "Lehrbuch der Allgemeinen Chemie" by Ostwald, chemical studies which he concluded "with some relief" a year later. We do not know where he learned the theory of relativity, although there is evidence that approximately at that time he had mastered Richardson's "Electron Theory of Matter", which expounds in fair detail whatever was known in 1916 concerning electromagnetic theory and atomic structure. A little later, probably in 1921, he read Weyl's "Raum, Zeit, Materie" and was impressed by the power and flexibility of variational methods which were used systematically in that book. He immediately assimilated the technique and the spirit of the method and proceeded to use it for new problems. Thus originated in his third University year his first published papers ($N^\circ 1$, 2).

Among Fermi's papers at the University of Chicago is a very interesting document referring to his early university years. It is a small leather-bound note book which he filled between July and September, 1919, and which contains a sort of physics vademecum. It is divided in parts and, as was his habit, is written in pencil almost without erasure.

This booklet gives us a clear picture of Fermi's scientific preparation and intellectual progress in 1919 when he was between 18 and 19 years old. The first 28 pages contain a summary of analytical dynamics and are dated Caorso, July 12, 1919. In it he develops the theory of Hamilton and Jacobi, reaching very advanced topics with extreme concision, but equal clarity. There are no indications of his sources of information, but very likely they are the works of Poincaré which he was studying around that time and also Appel.

Twenty-five pages on the electronic theory of matter follow; they are dated Rome, July 29, 1919, and they contain a résumé—as usual a very concise one—of the subject. In these pages he treats Lorentz's theory, spe-

---

(3) E. PERSICO, *loc. cit.*

cial relativity, black-body theory, diamagnetism and paramagnetism. For this part there is also a bibliography listing several of the most important books on this subject, including Richardson's "Electron Theory of Matter" which we know he studied carefully.

Bohr's first papers on the hydrogen atom are also mentioned, although at that time they were little known and appreciated in Italy. The following section of 19 pages is dated Rome, August 10, 1919, and contains in greater detail the black-body theory according to Planck. This part is followed by an extensive bibliography on radioactive substances and their decay taken from Rutherford's "Radioactive Substances and Their Radiations". There are no comments, and it is dated Caorso, September, 1919. The following chapter, from pages 81 to 90, is devoted to Boltzmann's H–theorem and kinetic theory and is dated Caorso, September 14, 1919. We find the usual succinct but clear exposition of theory with some applications. The method used for establishing the H–theorem is the same as that used by Boltzmann which involves a detailed analysis of all collisions. The booklet, totaling 102 pages, concludes with two bibliographies taken from Townsend's book on gas discharges. They deal with electrical properties of gases and photo-electricity. The last notes are dated Rome, September 29, 1919, and are followed by a table of contents.

This booklet shows many of the author's characteristics in an embryonic stage. The choice of material is made with surprising discrimination, especially if one considers the author's age and the fact that he was essentially self-taught. Another characteristic is that Fermi, although never repulsed or frightened by any mathematical difficulty, does not seek elegant mathematics for its own sake. Whether a theory is easy or difficult does not seem to concern him; the important point is whether or not it illuminates the essential physical content of the situation. If the theory is easy, so much the better, but if difficult mathematics is necessary, he is quickly resigned to it. One also notices an appreciable difference between the sections in which the logical structure of the subject predominates over its experimental content and those of a more empirical character. In the first, one perceives the master's hand; in the second, the lack of experience and of a critical evaluation of the many papers quoted is apparent. All told, it is surprising that after one year of university work a student should be able to put together such a booklet, which would be very creditable even for a teacher with a long educational career behind him.

The physicists with whom Fermi came in contact at Pisa are: Professor L. Puccianti, then Professor of physics and Director of the physics laboratory, Dr. G. Polvani, Puccianti's first assistant and two other assistants:

Drs. Pierucci and Ciccone. Puccianti was gifted with a keen mind but was a lazy person who, while still young, had done some interesting work in atomic spectroscopy. At the time Fermi went to Pisa, Puccianti had ceased doing original research, although he remained interested in a deep and critical understanding of classical physics. Fermi became a good friend of Polvani. Among Fermi's classmates we remember Rasetti, who was very close to Fermi, and Carrara, who also had a fellowship at the Scuola Normale Superiore.

Besides studying physics Fermi also acquired a vast and deep knowledge of mathematics. As a matter of fact, even though he occasionally showed scorn for certain parts of mathematics which he deemed too formal or too little imaginative or too critical, he certainly was not mathematically naive or ignorant of the most modern and subtle mathematical questions. If needed, Fermi could give the most rigorous proof of a theorem, and often, on an example, he would show a refined critical approach, abandoning it later for the sake of speed and simplicity. There is no doubt that from his studies at Pisa he derived a supply of mathematical facts, ideas and methods that he used for the rest of his life. Whenever he needed an ingenious and powerful mathematical method, Fermi always had it ready in some corner of his mind, even when it involved mathematical notions above and beyond the common knowledge of professional theoretical physicists.

The studies at the Scuola Normale Superiore proceeded brilliantly: he always obtained the highest possible grades, except in drawing and in some chemistry courses. This cost him no effort and left him, as we noted, ample leisure for his private studies.

He enjoyed enormous prestige among his fellow students and his superior ability was recognized by everybody; indeed, it was known that the Scuola Normale Superiore had an extraordinary man among its students.

The doctoral dissertation, by tradition, was on an experimental subject, and its content is essentially reproduced in papers N° 6 and 7. The day of the oral examination the classroom was full, and his colleagues expected a memorable performance. Instead, the examination, although excellent, was not exceptional. Anyway, he obtained his degree *cum laude* in July, 1922. Three days later Fermi passed also the examination required for the "Diploma" of the Scuola Normale with a thesis on probability (N° 38). On that occasion some of the mathematicians made numerous remarks and criticisms of his solution of a certain equation; however, again he obtained the maximum of grades, *cum laude.*

Immediately hereafter, Fermi returned to live with his parents in Rome and at that time became acquainted with Senator O. M. Corbino, Professor of experimental physics and Director of the Physics Institute at the University of Rome. Corbino, although already at that time considerably absorbed in politics and business, was the most open-minded among all Italian physicists in a position of authority. Gifted with a brilliant mind, a true love for science and a generous human approach, he exerted an influence on the development of physics in Italy far greater than that derived directly from his scientific contributions. As long as he lived, he remained the wise advisor, friend and protector of Fermi and of the whole group which gravitated around him. The generosity with which he gave this necessary help, and the obvious satisfaction and delight with which he followed first Fermi's successes and then those of the group which populated Rome's Institute, are the clear signs of his superior intellect and generous heart. To my knowledge there are no documents or direct evidence of the relations between Fermi and Corbino around 1922. The acquaintance, which had quickly grown into friendship, between Fermi and the three distinguished mathematicians Castelnuovo, Enriques and Levi-Civita dates back to the same time. These three scientists were all at that time on the Faculty of Sciences at the University of Rome and taught there. Although they were professional mathematicians, they followed with more interest than most Italian physicists the developments of theoretical physics, especially in the field of relativity. Fermi also met Professor Volterra, who was primarily interested in problems of classical mathematical physics, but did not become especially intimate with him. These cordial relations with the mathematicians are reflected in the lectures that Fermi held at the Mathematical Seminar of the University of Rome, several of which appeared in print in the "Periodico di Matematiche", directed by Enriques (N° 22, 29, 34, etc). At that time there was no physics seminar at Rome and research was much more active in pure mathematics than in physics.

During the winter of 1923, Fermi won a scholarship from the Ministero dell'Istruzione Pubblica to study abroad and chose to work with Max Born in Göttingen. The time was a critical period of incubation for the new quantum mechanics, and Göttingen was one of the major centers involved; in spite of this, Fermi did not profit very much from his stay there. It is not easy to understand why this happened: we may surmise that his love for concrete, well defined problems, and his distrust of questions that he considered too general and abstract, may have repelled him from the then current speculations. These were certainly somewhat nebulous, and even worse for him, commingled with philosophical overtones; nevertheless, later they were destined to bring about the new quantum mechanics. The preference

shown by Fermi for concrete problems especially in his early years, is probably to be traced to his cultural formation, and to the fact that he was practically completely self-taught. Concrete problems probably gave him the
immediate sensation of the importance and a check of the correctness of
his work, whereas general and abstract questions were much more difficult
to plumb, and moreover isolation was a great obstacle in assessing the value
of the results obtained. In later years Fermi considerably changed his objects of research and was less inclined to confine himself to detailed ingenious investigations of concrete problems.

From a personal point of view, it is barely possible that during the Göttingen period the physicists in his age group, such as Heisenberg, Pauli, Jordan and others, all men of exceptional talents, who should have been his
companions, may not have recognized his ability and quite unintentionally
left him out of their intellectual community.

Anyway, Fermi remembered the following months, when he moved to
Leyden with a Rockefeller Fellowship to work in P. Ehrenfest's Institute, as
more fruitful. Ehrenfest was one of the greatest teachers of physics and a
man of warm human interests. Very early he recognized Fermi's ability and
encouraged the timid young Italian. Anyone who met Fermi in later years
might be surprised perhaps that he ever needed encouragement, but Fermi
himself often mentioned the fact. While he was aware of his superiority
over the physicists he had met in Italy, he said he did not have any way to
compare his capabilities with those of the great international names. Ehrenfest, who personally knew most of the international figures, gave him just
such a standard of comparison. The interests prevailing among Dutch physicists at the time are reflected in some of Fermi's papers (N° 21). And on
the other hand some eyewitnesses, such as Uhlenbeck, relate that Ehrenfest, who had delved deeply in statistical mechanics, was impressed by Fermi's ideas on the ergodic theorem (N° 11).

On the whole, the time spent abroad does not seem to have been of decisive importance for Fermi's development, as it might have been had the
circumstances in Göttingen proved more favorable. Fermi was accustomed
to intellectual isolation and to learning from books and journals more than
from personal discussions (a habit which he reversed in later years). He returned to Italy without having absorbed either the "spirit of Copenhagen"
or that of Göttingen. Heisenberg's great papers on matrix mechanics of
1925 did not appear sufficiently clear to Fermi, who reached a full understanding of quantum mechanics only later through Schrödinger's wave mechanics. I want to emphasize that this attitude of Fermi was certainly not

due to the mathematical difficulty and novelty of matrix algebra; for him, such difficulties were minor obstacles; it was rather the physical ideas underlying these papers which were alien to him.

When he returned to Italy in 1924, he began teaching an introductory mathematical course for chemists at the University of Rome. In 1925 he obtained the "Libera Docenza" and moved to the University of Florence to teach mechanics and mathematical physics. The physics laboratory, directed by Prof. Garbasso, was located on the Arcetri hills near the villa where Galileo spent his last years. In Florence he found Rasetti, who was at the time Garbasso's assistant. In February, 1926, Fermi took part in the competition for a chair of Mathematical Physics at the University of Cagliari in Sardinia. The referees were Professors Levi-Civita, Volterra, Somigliana, Marcolongo and Guglielmo. Whereas the first two favored Fermi, the majority chose for the first place Professor G. Giorgi, and Fermi remained with his temporary appointment in Florence.

Fermi was disappointed by the result of the competition, which he considered unjust; moreover, he naturally wanted to improve his modest financial conditions and was impatient to see his ability recognized officially.

Fermi's early correspondence with some of his closest friends, such as Persico, shows evidence that he was worried about his career. I mention this fact because later Fermi seemed to consider such preoccupations on the part of young scientists as something unnatural or exaggerated or strange. Probably once he had overcome the difficulties connected with the beginning of his own career, he took a more detached and lofty attitude towards these human vicissitudes, but at an earlier time he had certainly been anxious.

The Florentine period saw Fermi's first major contribution to theoretical physics: the discovery of the statistical laws which govern particles subject to Pauli's exclusion principle, or fermions as they are called today (N° 30). Soon after the discovery of this statistics, Corbino succeeded in having a chair in theoretical physics, the first in Italy, established in Rome and the competition for it was judged in November, 1926, by Professors Maggi, Cantone, Garbasso, Majorana and Corbino. They unanimously agreed that Fermi was the outstanding candidate, saying that "they felt they could rest on him the best hopes for the affirmation and future development of theoretical physics in Italy". The other two successful candidates were his childhood friend, Prof. Persico, and Prof. Pontremoli. Fermi's actual appointment to Rome was opposed by an older member of the physics faculty, but Corbino readily overcame the obstacle. Persico replaced Fermi at Florence and in the fall of 1926 Fermi moved to Rome. In the Institute at Via Panisperna 89[a] he started one of the most fruitful periods of his scientific career. The decade from the discovery of his statistics to 1936 was probably the golden age of Fermi's life.

The old physics building in Via Panisperna, although built around 1880, was still perfectly adequate for scientific work at that time and compared favorably with other major European laboratories. The equipment was fair and mainly included instruments for optical spectroscopy with good modern Hilger spectographs and adequate subsidiary apparatus. The shop was old-fashioned with rather poor machines; the library, on the other hand, was excellent. The location of the Institute, surrounded as it was by a small park on a hill in a rather central part of Rome, was convenient and beautiful at the same time. The gardens landscaped with several palms and bamboo thickets, the silence prevailing, except at dusk when many sparrows populated the greenery, made it a most peaceful and attractive center of study. I believe that everybody who worked there until 1937, when the physics department moved to the new "Città Universitaria", has still in his heart an affectionate and poetic feeling for the old place. The upper floor of the building was occupied by Corbino's residence; the first floor contained the research laboratories and the offices of Corbino, Lo Surdo and Fermi, as well as the library; the ground floor contained the shop, the classrooms and some laboratories for the students; the basement contained the electric generators and other facilities. Fermi, with Rasetti, and later with their students occupied the whole south side of the first floor; Lo Surdo most of its north side. As time went on, however, the number of people associated with Fermi increased so much that they filled most of the first floor. Neighboring quarters were occupied by Prof. G. C. Trabacchi, who was the chief physicist at the Health Department (Sanità Pubblica). He had an excellent supply of instruments and materials which he generously loaned whenever we needed them. This earned him the nickname of "Divine Providence".

Corbino, who used to spend part of the morning in his study, often visited Fermi's quarters and stopped to talk about physics and other subjects with him and his students.

When Fermi arrived in Rome he must have felt a strong urge to find an adequate scientific environment. There was then in Rome, as mentioned above, a group of eminent mathematicians considerably older than Fermi, such as Volterra, Levi-Civita, Castelnuovo, Enriques and others. Fermi was especially friendly with the last three, but the excellent personal relations and the high opinion which they had of each other were not sufficient to create intellectual exchanges; these were hindered by the essentially different points of view, scientific interests, and the considerable age difference. Among the physicists, Corbino was the only one open-minded enough and with sufficient preparation to follow and appreciate the developments of modern physics, the rapid progress of theoretical physics in general, and especially of quantum theory. However, he was busy with his political and business activities and, while he was fully aware of the necessity of rejuvenating physics in Italy and of introducing new young workers to the field, his direct scientific activity was limited. On the other hand, his political and administrative help and the enthusiasm with which he fostered the development of a physics school in Rome were of paramount importance.

The first step following the arrival of Fermi in Rome was Rasetti's transfer from Florence to Rome. He became Corbino's first assistant (Aiuto) and immediately started to try to revive experimental work. At this point I might introduce some personal recollections in order to illustrate the formation of the physics school in Rome. In 1927 I met Rasetti through a friend and a schoolmate of mine, G. Enriques, the son of the mathematician. While we went mountain climbing together, I learned from Rasetti several physical theories; but even more important, I gained a clear impression that physics in Rome was awakening. As a matter of fact, in 1925 I had already heard some lectures by Fermi at the Mathematics Seminar of the University. They had made a deep impression on me because they showed me for the first time a young man who was deeply versed in subjects of which I had barely caught a glimpse (by reading the quantum theory book by Reiche) but had never met in my regular university studies. However I had not yet met Fermi personally and I did not have any way of comparing the young lecturer with the famous men I knew only through my readings.

During the summer of 1927, through Rasetti and G. Enriques, I came to know Fermi personally and I immediately had the sensation of having found an exceptional teacher. Fermi himself, in conversations, on hikes and at the seashore, asked me several simple questions in mathematics and classical physics, perhaps to discreetly test my knowledge and ability. In September, 1927, on my return from an expedition to the Alps where I had been with Rasetti, G. Enriques and others, I went to the International Physics Conference in Como and there had the unforgettable experience of seeing in person the great physicists whose names I had read in books: Lorentz, Rutherford, Planck, Bohr and many others, and then a group of extremely young men: Heisenberg, Pauli and Fermi. It was clear that, scientifically at least, Italy was represented by Fermi. During the short period of the Conference I also learned a considerable amount of physics because

Rasetti and Fermi would point out to me the various celebrities and at the same time tell me their most important achievements. I returned to Rome in the fall of 1927; decided to follow my old desire of studying physics, and within a few months left the engineering studies which I had been pursuing until then.

On that occasion for the first time I saw Corbino's far-reaching influence. He overcame with ease the far from trivial administrative difficulties in order to save me the loss of credit for one year of engineering studies. Without anybody taking special notice of it, I had become Fermi's first pupil, at least in a formal sense. The school in Rome had been started.

In June, 1927, without my having heard anything about it, Corbino had made an appeal to one of the undergraduate physics classes for engineers to enlist some new physics students: Edoardo Amaldi, the son of the mathematician Prof. Ugo Amaldi, followed this invitation, and thus he also entered the small group of students who were to be personally tutored by Fermi with the hope of developing among them research physicists and possible collaborators. Perhaps later this would lead, eventually, to a renaissance of physics in Italy. However this goal was not explicitly formulated at the time.

After a few months of study I talked to my friend and schoolmate at the engineering school, Ettore Majorana, and he also joined our group. Majorana, in intellectual power, depth and extent of knowledge, was much above his new companions and in some respects—for instance as a pure mathematician—superior even to Fermi. His exceedingly original intellect and, unfortunately, also his natural pessimism and exaggerated self criticism, drove him to work alone and to lead a solitary life. On the whole he did not participate very much in our work, and he limited himself to helping us in difficulties or to stunning us with his original and novel ideas and methods, or with his ability as a lightning mental calculator. Later he isolated himself even more, and by 1935 he had disappeared from the University, seldom leaving his private residence.

Other students frequented the Institute and once in a while attended Fermi's private lectures.

These were completely improvised and informal. In the late afternoon we met in his office and often conversational topics gave rise to a lecture: for instance, we would ask what was known about capillarity and Fermi would improvise a beautiful lecture on the mathematical physics of capillarity. In this fashion we reviewed many subjects at an intermediate level, more or less corresponding to a beginning graduate course in an American university or to the famous "Introduction to Theoretical Physics" by Planck or, in English, to the books by Slater and Frank. Sometimes, however, the level became higher and Fermi would explain to us a paper which he had just read: in this way we became conversant with the famous papers by Schrödinger on wave mechanics. We never had a regular "course". If there was an entire field of which we knew nothing and we asked Fermi, he would limit himself to mentioning a good book to read. Thus when I asked for some instruction in thermodynamics, he told me to read the book by Planck. However, the readings he suggested were not always the happiest ones because he probably mentioned only the books which he had studied himself, and these were not necessarily the best suited from a pedagogical angle. After his lecture, we would write down notes on it, solve (or try hard to) the problems which he had given us, or others which we ourselves invented. The rest of our time we spent on experimental work.

The instruction was chiefly in theoretical physics and no distinction was made between future theoreticians or experimentalists. Fermi himself, who at the time worked mainly in theory, was also interested in experimental work. His knowledge and interests embraced all physics, and he diligently read several journals. He preferred concrete problems and distrusted theories which were too abstract or general, but any specific problem, in any field of physics: classical mechanics, spectroscopy, thermodynamics, solid state theory and several more, fascinated him and was a challenge to his ingenuity and physical sense. Often just talking to him one saw a beautiful explanation develop, simple and clear, that resolved some puzzling phenomenon.

At that time we repeatedly had occasion to witness the execution of a new and original piece of work. Naturally it is impossible to say how much preliminary work Fermi had done consciously or unconsciously. Certainly there were no written notes. What one saw was the development of a calculation at a moderate speed, but with exceedingly few errors, false starts or changes of direction. The work proceeded almost as in a lecture, although more slowly, and at the end the manuscript, or at least the equations, were ready to be copied for publication with little need of improvements. A curious characteristic of Fermi's working habit was the steady pace at which he proceeded. If there were easy passages he still proceeded quite slowly, and a simple-minded observer might have asked why he wasted so much time on such simple algebra; however, when difficulties arose which would have stopped a man of lesser ability for who knows how long, Fermi solved them without a change of speed. One had the impression that Fermi was a steamroller that moved slowly but knew no obstacles. The final result was always clear and often one was tempted to ask why it had not been found long ago since everything was so simple and natural. Once used, a method was stored in his memory and often adapted to problems which appeared to be very different from the one which first originated the physical idea or mathematical technique. For instance, it is interesting to note the evolution and successive applications of the "scattering length" in papers N° 95 and 107, or the recurrence of the statistical theme applied to both atoms and nuclei.

Already at that time (1928) Fermi made little use of books: Laska's collection of mathematical formulae and Landolt–Börnstein's tables of physical constants were almost the only reference books he had in his office. If he needed some complicated equation to be found in a book in the library, Fermi would often propose a wager, saying that he would derive the equation faster than we could find it in a book. Usually he won. The only treatise that I know he read after he came to Rome was Weyl's "Gruppentheorie und Quantenmechanik".

The speed at which it was possible to form a young physicist at that school was incredible. Naturally, a good deal of the success was due to the immense enthusiasm aroused in the young people, never by exhortations or sermons, but by the eloquence of Fermi's personal example. After some time spent at the Institute in Via Panisperna, one became completely absorbed in physics, and in saying "completely" I am not exaggerating.

Fermi did not like to assign subjects for doctoral dissertations, or in general to suggest subjects of investigation. He expected the students to find one by themselves or to obtain one from some colleague who was more advanced in his studies. The reason for this, as he later told me, was that he did not easily find subjects simple enough for beginners: he generally thought of problems that interested him personally and were too difficult for students. Rasetti was very generous in teaching the experimental techniques that he knew or in loaning apparatus that he no longer used, but it was difficult to work with him because he had peculiar idiosyncrasies and irregular schedules of work. A strong and long-lasting personal friendship developed among all the participants in this adventure. Age differences were small: Fermi, the oldest, was 26 years old in 1927; Amaldi, the youngest, was 19. Corbino occasionally came to the evening lectures, but the event was rare. However, he took a strong interest in the welfare of the group, in the career problems and relations of the young men with the rest of the world. As was to be expected, word of what was happening in Rome spread quickly among the young Italian physicists, and soon we started to receive visits from G. Gentile Jr., B. Rossi, G. Bernardini, G. Racah, G. C. Wick, and later we had for longer periods L. Pincherle, R. Einaudi, E. Fubini, U. Fano and many others. By 1929, it became clear that whereas the theoretical situation was well in hand, it was necessary to strengthen our experimental activities. To import new experimental techniques to Rome, members of the group had to work in different laboratories to learn them on the spot. Thus Rasetti went to Pasadena, to Millikan's laboratory where he did important work on the Raman effect; I went to Amsterdam to Zeeman's laboratory to study forbidden spectral lines and Amaldi went to Debye's laboratory in Leipzig where he worked on X-ray diffraction of liquids. Later we supplemented these visits, which at first had the purpose of using facilities not available in Rome, for the completion of problems already started there, by trying entirely new fields. In this connection, Rasetti spent some time at Dahlem in Lise Meitner's laboratory, and I worked in Otto Stern's laboratory in Hamburg. The plan was successful, and without these periods of experimental training it would have been impossible later to perform the complex neutron work rapidly and efficiently. However, even from abroad, we kept in close contact with the group in Rome, and either by letter or during vacations we discussed theoretical problems with each other. There are many traces of this exchange in papers by Fermi, Majorana, Wick and others. During this period Fermi went abroad only for short visits. He was by now accustomed to being rather isolated intellectually because only Majorana (and he was rather inaccessible as mentioned above) could speak with him about theory on an equal footing. On the other hand, about that time, the first victims of Nazi persecution, barred from their native countries and attracted by Fermi, began to arrive in Rome. Bethe, Placzek, Bloch, Peierls, Nordheim, London and others spent some time in Rome, often on their way to the United States. Among the Americans, we had a visit by Feenberg, whom Majorana especially liked. This mutual attraction manifested itself by their sitting in the library facing each other in silence because they knew no common language.

Life in Via Panisperna was very methodical. One worked from about 9 a.m. to 12:30 p.m. and from 3 p.m. to 7 p.m. Of course this schedule was self imposed and everybody was free to keep it as he wished. Work in the evening was practically unknown and on Sundays we often went for a hike in the vicinity of Rome or for a mountain trip. During the winter there was always some skiing expedition, and during the summer, a trip abroad or a vacation in the Alps. Already in 1930 Fermi had gone as far as Ann Arbor (Michigan), attracted there by its brilliant summer school of theoretical physics and by his Dutch friends Uhlenbeck and Goudsmit who had moved there.

The most significant personal events of this period were Fermi's marriage to Laura Capon, in July 1928, and his appointment by Mussolini to the Italian Academy, in 1929. This honor, though well deserved, was unexpected because Fermi's reputation at the time was limited to physicists and, according to tradition, at his age, academic honors were not yet due. The appointment came probably at the instigation of Corbino, who, although not a Fascist, had been a member of one of the early cabinets of Mussolini. This event changed Fermi's financial condition and had a beneficial effect on the subsequent development of physics in Italy, since its representative in the National Academy was undoubtedly the most qualified man available. However, even after his appointment to the Academy, Fermi wielded relatively little political influence because he was unwilling to sacrifice any time to occupations outside of physics and he did not like to participate in administrative matters or political affairs. He was able to resign his post at the "Enciclopedia Italiana Treccani" (see e.g. N° 83) and at the National Research Council, which he had taken on his arrival in Rome, mainly because of his financial position. He made some desultory attempts at increasing research facilities by the creation of new jobs and subsidies, but these attempts did not go farther than obtaining Mussolini's approval on a memorandum which was later pigeonholed. It took many more years before·the younger generation could prevail in the important matter of university appointments.

In the meantime, the most important scientific event of those years, the formation of quantum mechanics, had taken place without important contributions from Italy, at least in the establishment of the principles, but Fermi had contributed to its applications in an important way. Fermi's statistics was born independently of quantum mechanics and certainly before Fermi himself had mastered it completely. In the original papers we can trace this effort to clarify and assimilate quantum mechanics (N° 37, 39, 42) between 1926 and 1931. Schrödinger's papers were the first to be understood and aroused great interest and enthusiasm. Fermi explained them quickly to his friends; later to Corbino, who remained skeptical for some time; and still later he spoke on the subject to the Mathematics Seminar, where the professional mathematicians, older in age and less familiar with the experimental background of physics, raised several ingenious objections to the interpretation of quantum mechanics commonly accepted.

Thus, for instance, paper N° 59 originated from a discussion in which Prof. Castelnuovo raised many questions. Fermi was inclined to be somewhat impatient with people who did not understand the new developments of quantum mechanics, but naturally he treated very differently silly objections, of which there were many, and genuine difficulties, like those Castelnuovo raised.

Once in a while he complained that even persons for whom he had the highest respect and admiration, such as Corbino, would occasionally show a skepticism towards quantum mechanics and its interpretation which he deemed based on lack of understanding. It must be said, however, that in his last years Fermi seemed less convinced that the current interpretation of quantum mechanics is the last word on the subject. This resistance against quantum theory was shown mainly by people older than Fermi and his group, because the younger physicists either understood or believed the new theories, and in any case learned to use them, even if they had not completely assimilated them.

The advent of quantum mechanics in Fermi's opinion, and in Corbino's also, signaled the completion of atomic physics. The fundamental questions were solved, and the future lay in the direction of the exploration of the nucleus, or of complicated structures leading ultimately to biology.

This represented a radical change in the research projects of the Institute. We must remember that our experimental tradition in Rome went back to the spectroscopic work of Puccianti in Pisa: all our successes obtained up to that time in experimental physics were on spectroscopic subjects; the available equipment was spectroscopic; and our knowledge was mainly in the field of atomic physics. Rutherford and the work of his school were rather alien to us.

Thus the change to nuclear subjects cost us considerable effort. It was not a whim or a desire to follow a fashion, but was the result of a deliberate plan which we debated with plenty of vigor. The first step towards its realization was Rasetti's trip to Dahlem to learn nuclear techniques. A nuclear physics conference at Rome in 1931, organized by the Italian Academy, helped to familiarize us with current problems, and we changed the subject of our readings to matters connected with the nucleus. Fermi, later in his career, at the end of the Second World War, experienced a similar change in his field of work when he left neutron physics, in which he was recognized as the greatest authority, to study high energy physics.

The first work on a nuclear subject is the review article N° 72, and the first great success was the beta ray theory (N° 76), where we find the coalescence of previous work on radiation theory with Pauli's idea of the neutrino. A little later, artificial radioactivity research was in full swing. As the history of that period is told in detail in the introductions to papers N° 78,

84 to 110 and 112 to 119, it is omitted here. The neutron work absorbed Fermi completely for the remainder of the time he was to spend in Italy.

Unfortunately, the political horizon clouded with ominous signs. Since 1936, approximately, at the Institute we had a feeling of impending catastrophe. The days of serene, undisturbed work were gone forever. From 1930 to 1936, during a period of relative exterior quiet, we have the peak of Fermi's activity and the prime of the school in Rome. The history of this period is recounted in the introductions to the papers of those years. There is little to add. The neutron work was carried out with the kind of extreme rapidity that was possible only for a small group working in complete harmony and without administrative encumbrances. In this connection it is worth mentioning the low cost of this research. The Consiglio Nazionale delle Ricerche, on recommendation of its secretary, G. Magrini, immediately appropriated 20,000 lire (1000 U.S.A. dollars at that time) which could be spent with complete freedom. This was our only subsidy, apart from our regular salaries. The neutron work also had an untoward consequence: it became absolutely impossible for Fermi, because of lack of time, to continue his extracurricular teaching of promising young men; neither was he able to spare time for foreign visitors; from 1934 to 1938, first they became rare, then practically disappeared.

The Ethiopian war marked the beginning of the decline of the work at the Institute, and the death of Prof. Corbino on January 23, 1937, brought further serious complications. Prof. Lo Surdo was appointed director and he did not understand or appreciate the merit of Fermi's work. The Institute itself was moved to the new university campus, to a larger and more modern building; however, this move cost invaluable time. All these events and the grave political situation materially hampered Fermi's work. Finally, the Fascist racial laws of 1938 directly affected his wife and possibly his two children. This fact, and probably his deep, although mute resentment against an injustice that offended his sense of fairness, were the final arguments that convinced him to leave Italy. He was not to return to his native country again until 1949.

Already in the past Fermi had had offers of permanent positions from the Eidgenössische Technische Hochschule in Zürich and from several American universities. Some of these offers included not only substantial economic advantages, but also the availability of research facilities far superior to those in Italy. There is no doubt that Fermi considered these offers very carefully and several times was seriously tempted to accept them, but for various reasons he ultimately decided against them. The events of 1938, however, gave him a final impetus, and the award of the Nobel prize in Stockholm gave him a pretext to leave Italy while avoiding possible political reprisals. From Stockholm he travelled directly to New York, and thus began a second period of his life.

"Lo primo tuo rifugio e 'l primo ostello" (Thy first refuge and first home —Dante, Divina Commedia, Paradiso XVII, 70) was Columbia University. Fermi might have described Columbia in these words, quoting Dante as he often did to show off, in a light fashion, among his Italian friends.

Indeed, one cannot properly speak of a refuge, because as soon as the rumor spread that Fermi was in the U.S.A. to stay, several universities made him excellent offers. At that time American universities had suddenly found a number of distinguished European scholars who had been dismissed from their positions at home and were looking for new positions. Hitler's persecutions and, on a smaller scale, Mussolini's, had forced many scientists, from famous ones to young men just beginning their careers, to leave their countries, and England and America with a liberal and generous sense of hospitality and an intelligent understanding of their own long-range interests, performed a good part of the rescue operations. This was an undertaking which, in a bleak period of human history, when the blackest and most hideous crimes recorded in history were committed, honors the scientific community and shows that even at a time of bestiality, mankind is open to noble deeds.

Fermi, however, was in many ways spared the difficulties of that time. He was sufficiently famous to find a suitable position without difficulty and he was young enough to adapt easily and rapidly to the new surroundings. This adjustment was rendered even easier by his previous acquaintance with the United States and by his fondness for the American way of life. At Columbia University he had personal acquaintances; he himself, Rasetti, Segrè and Amaldi had worked there. Besides the excellent local instruments, he also found copies of some of the instruments he had used in Rome; moreover there was a cyclotron. The study of neutrons had been one of the major subjects of investigation at Columbia and it would be easy to resume with larger means the work interrupted at Rome. Last but not least the head of the Physics Department, Professor G. B. Pegram, truly a Southern gentleman, had taken a personal interest in the Rome group and a warm friendship developed between him and his Roman guests. Thus there were good reasons for Fermi's choice of Columbia. As soon as he was settled, he tried to form a new group of co-workers. H. L. Anderson was among his first students there. However, neutron work in the style and tradition of Rome was nearing its end. Without the discovery of fission which suddenly changed the situation, and without the outbreak of the war, it is probable that Fermi would have left neutron physics and turned to another subject. Instead, the discovery of Hahn and Strassmann suddenly opened new horizons to nuclear physics and especially to neutronology. At the time, Fermi was without doubt the greatest expert on neutrons. His intuitive feeling for their behaviour was like that of a radio expert for the behaviour of circuits. He did not need to calculate in order to predict the results of an experiment, but true to his nature, after each experiment he performed numerous and detailed calculations which he saved and classified in an extremely orderly fashion. In this way he developed and increased a "thesaurus" of data, a mixture of theoretical and experimental results which became a most valuable and unique tool for his further work. He used to call the container of these sheets the "artificial memory". The artificial memory started as a large envelope, later grew into a desk drawer, and finally filled an entire filing cabinet.

As soon as the announcement of the discovery of fission reached Fermi, he immediately saw the possibility of the emission of secondary neutrons and of a chain reaction, and at once he started experimenting with great alacrity in this direction. Nothing tells the story of the experiments which led to the pile better than the papers contained in the second volume of these Collected Papers. Most of the papers were unpublished and certainly not intended for publication in their present form; on the other hand, they represent a scientific chronicle of enormous interest, and taken together with their introductions, they give an exact account of the events of that period so important for both science and history. Anderson, who worked with Fermi from 1938 on and collaborated in a large portion of the experiments which led to the chain reaction, gives us the eyewitness report. I saw Fermi repeatedly during that period and we worked together again briefly during one of his visits to Berkeley, but the secrecy requirements of the time unfortunately prevented a free exchange of ideas. This, however, did not prevent us from discussing future projects. For instance, during a visit of mine to New York towards the end of 1939, while we were on a walk, along the Hudson River, near Leonia where he lived, we discussed plans for the research that was to lead to the discovery of Pu^{239} and to its use as a nuclear fuel.

During the last years of his stay in Italy, Fermi had developed a certain reticence. As far as I know, there were no special reasons for it at the time, but understandably it became extreme under the conditions of war work. Among the many scrupulous observers of security rules in atomic energy work, before and after the institution of the Manhattan Project, I doubt that there was anybody more discreet than Fermi. The legal status of "enemy alien" in which Fermi found himself also produced many anomalous situations, most of which were easily solved by common sense and the relative flexibility of the administrative and bureaucratic systems then prevailing. Some of these episodes were rather funny and did not leave any resentment on either side; mostly they were sources of amusing stories to be told among friends.

The chain reaction project was begun with a completely inadequate staff and equipment. Perhaps Fermi thought he might be able to repeat, on a somewhat larger scale, a work similar to the neutron research in Rome. He certainly did not realize, as very few of the scientists did, the colossal proportions in manpower and means that the project was to assume before its successful completion. Fermi was always reluctant to take administrative responsibilities so that in the Manhattan Project too, at least officially, he never became a prominent personality from the administrative or political point of view. Scientifically, he was undoubtedly the brightest star, and this was universally acknowledged. Because of his unparalleled knowledge of neutrons, he was consulted whenever serious difficulties arose. The sequence of researches directed towards the achievement of the chain reaction is a thing of beauty: the logical order of the experiments, the evaluation of the results and their use to direct the next step are to be followed in these Papers. Whoever has any scientific experience will be deeply impressed by a perusal of them. We find here a sequence of investigations connected with each other by an iron logic and a keen intuition; they stand equal to some of the major classical works of experimental science.

In the fall of 1942, during a visit to Chicago, Fermi locked me in a room with some of the reports (including numerical data and connected experimental results) on the exponential experiments and the theory of the pile.

I still remember how after a couple of hours of study I was dumbfounded by the success already reached and by the methods followed. Having read those reports, I saw clearly that a chain reaction with natural uranium was on the verge of being realized.

Indeed, on December 2, 1942, the pile under the Stadium of the University of Chicago became critical. This brought to an end another cycle of investigations. Four years had elapsed between this event and the discovery of fission, seven had passed since the discovery of slow neutrons, ten since the discovery of the neutron. Such was the unprecedented speed of this work.

Even in this period so overloaded with work, Fermi did not completely neglect his teaching activity. Instead of teaching classes or his private seminars, as in the past, he taught his co-workers during the work and many of them still remember with pride and pleasure the period of apprenticeship under Fermi's guidance. Once he had reached the main goal of the Chicago Project, Fermi typically enough, left to others the further development work in order to start something new and to lend his help where the diffi-culties were the greatest: at the time at Los Alamos.

The assignment of that laboratory, ably directed by J. R. Oppenheimer, was to make an atomic bomb with either U^{235} or Pu^{239}, as soon as they were delivered by other branches of the Project. The difficulties of this as-signment were far from trivial; actually, at the beginning, nobody knew how to build an atomic bomb at all. Fermi did not have any specific scien-tific duty nor was he assigned any particular administrative responsibility; he was on the governing board of the laboratory with which the director consulted on all important matters, but otherwise he was a sort of oracle whose job was to solve problems above the ordinary capabilities of the staff, distinguished as it was. J. von Neumann, as a consultant, had a somewhat similar job.

Among the various activities at Los Alamos, Fermi took a direct inter-est only in the water boiler, a homogeneous reactor which had been built there; but in general he participated in all novel or unusual problems. I re-member having listened in his office to discussions on hydrodynamics with von Neumann. They took the strange form of a competition in front of a blackboard to see who could first solve the problem; von Neumann, using his insuperable lightning analytical skill, usually won. Occasionally, such discussions could be interrupted by unexpected events. For instance, dur-ing one of these hydrodynamical sessions I once witnessed the arrival of a first class electronics expert who was confronted with a new and very dif-ficult problem in circuitry. Within about twenty minutes Fermi concocted a circuit which would have solved the problem, but no one knew whether a tube with the necessary special characteristics even existed. Consultation of a tube manual disclosed that the required tube was really available and the apparatus was promptly built and worked satisfactorily.

The time in which the first atomic bomb would be ready for a test at Alamogordo was approaching. Several times we went to the New Mexico desert for preparations. Finally, shortly before daybreak on July 16, 1945, after a stormy night, we saw the light of the first atomic bomb. From a dis-tance of about ten miles, through very dark glasses, we saw the blinding light of the explosion, an imposing and awful sight. The problem of mea-suring the energy release was immediately although crudely solved by Fermi. With the help of some confetti which he dropped, he measured the displacement produced when the front of the shock wave reached his observation point. Within a few hours, Fermi went with a tank to the crater of the explosion in order to measure the radioactivity of the sand.

That explosion signified in a way the end of the initial phase of the Los Alamos project. The major technical assignment had been accomplished, too late to influence the war in Europe in the decisive way that had been planned. Germany capitulated on the very day on which at Alamogordo we exploded a tremendous pile of ordinary explosives in preparation for some tests to be performed on the atomic bomb, which was expected to be ready in a couple of months. President Roosevelt had died suddenly in April and could not see the end of the enterprise he had so effectively fostered.

The success of the Alamogordo test was promptly communicated to President Truman, then at Potsdam in conference with Churchill and Stalin. The grave decision on the use of the bomb rested with the President. As usual, for serious policy decisions in the U.S., even before the success of the Alamogordo experiment, the President had appointed a restricted advisory committee, called the Interim Committee to counsel him on the employment of the bomb and on other matters pertaining to nuclear energy. Its recommendations had a purely advisory value because the final decision, according to the U.S. constitution, rested with the President as Commander in Chief. Earlier, several scientists who had been very active in the Manhattan Project, wrote eloquent pleas to the President recommending against the military use of the atomic bomb. Among them I might mention James Franck and Leo Szilard, who were particularly worried by the war use of atomic energy. Naturally, all this occurred in secrecy, without public knowledge. The Interim Committee appointed as scientific consultants A. H. Compton, Fermi, E. O. Lawrence and J. R. Oppenheimer. In a meeting held at Los Alamos about June 15, 1945, after long and painful deliberations, they recommended the military use of the bomb. This happened at Hiroshima and Nagasaki.

At the beginning of August, Japan surrendered, and thus ended the Second World War. The scientists at Los Alamos once again started to think about peacetime research, and there were many projects floating in the air. During the fall, plans for the future were taking shape. Fermi was still on a leave of absence from Columbia University, but just at that time the University of Chicago planned to form a nuclear institute (the present Enrico Fermi Institute for Nuclear Studies) and repeatedly offered Fermi a position as its director. He absolutely refused a job that would have given him too much administrative work; fortunately Professor S. K. Allison, a distinguished scientist, a very able administrator, and a good friend of Fermi was appointed director. He supervised all the activities in the initial phases of the Institute and kept the job until after Fermi's death.

During the clear autumn days of Los Alamos, with the pressure of work relaxed, we resumed our old pastime of taking long walks, especially attractive in those beautiful and wild surroundings. Plans for the future were frequently our subject of conversation, sometimes in earnest, sometimes in a lighter vein. Among those that Fermi mentioned frequently was that in his mature years he would retire to a small college, with teaching duties only, and then write a book dealing exclusively with those theorems which are reputedly "well known" or "easily shown" and for that reason never really proved. He even began list of such topics, but unfortunately the book was never written. For the immediate future, while the new, extraordinarily powerful neutron sources could have attracted Fermi to the exploitation of these new technical means in the neutron field where he was—as said before—the foremost world authority, he quoted instead, jokingly, a motto which he attributed to Mussolini, although probably it originated with D'Annunzio: "O rinnovarsi o perire" (Renew oneself or perish), and said that henceforth one had to turn to new chapters of physics in which the future lay. True to his word, when he returned to Chicago he worked on neutrons for a relatively short time, long enough to obtain brilliant results which themselves became starting points for a long series of investigations, but he immediately started to prepare himself for the new meson physics which was just beginning to develop.

In the last months of his stay at Los Alamos he also started to develop a keen interest in electronic calculating machines. In his Roman days, he had used a small mechanical hand calculator to compute numerically the ψ's (cf. paper N° 43) and in order to perform several other applications of the Thomas-Fermi method of calculating atomic properties. Numerical problems and numerical analysis were completely familiar to him and he recognized at once that the new powerful electronic computers, strongly advocated by von Neumann, were opening up new and unpredictable possibilities. The powerful computers at Los Alamos always had a strong attraction for him, and he experimented and worked with them for several summers. The problems connected with nuclear fusion had also interested him deeply. Already in 1946, he had reviewed them in a lecture course at Los Alamos in which he expounded many novel and original ideas.

On his return to Chicago, while waiting for the synchrocyclotron and the new Institute to be finished, he again worked with neutrons at the Argonne Laboratory, using the pile as a source. We see in this period the conclusion of the investigations on neutron diffraction and on the scattering length, the origin of which can again be traced back to the Roman period. At that time he also sought to re-establish a system of instruction somewhat similar to that used in Rome. He had a new group of young pupils, most of them returning from Los Alamos, where they had worked in the theoretical group of Bethe, in the experimental group of Segrè, or in other groups. We shall mention among them Agnew, Chamberlain, Chew, Goldberger, Rosenfeld, Woods, and somewhat later, a new arrival from China, Yang. At Chicago Fermi took a very active part in all seminars and in many discussions; often, with a single remark, he sowed the seeds of further important developments. For instance, Maria Mayer in her classical paper on the shell model ("Phys. Rev." 75, 1969 (1949)) generously acknowledges: "Thanks are due to Enrico Fermi for the remark 'Is there any indication of spin-orbit coupling?' which is the origin of this paper". In the meantime meson physics was developing and Fermi immediately recognized the importance of the Conversi-Pancini-Piccioni experiment which he was the first to interpret correctly. But by now the Chicago synchrocyclotron was about to work and in Berkeley it had already been shown that such a machine was a powerful artificial source of mesons. Thus we enter the last phase of Fermi's experimental work; this is a series of important investigations on the pion-nucleon interaction performed with Anderson and younger pupils, such as Rosenfeld, Orear, Yodh and others. It was at this time that they coined the new words, pion and muon, for π-meson and μ-meson.

In this last period, Fermi, who by now was no longer fond of long trips, spent his summer vacations at Los Alamos or at Brookhaven, a most welcome guest of these Laboratories, especially of the first, to which he was particularly attached. He also spent a summer in Berkeley and he attended all the High-Energy Conferences in Rochester. Wherever there were new physics problems and young physicists you were likely to see Fermi arrive, always ready with some new and fruitful idea, and also ready to receive information, challenge and inspiration from his younger colleagues. In this period we also note a change in Fermi's methods of keeping up with scientific developments. He read less and less and relied more and more on conversation and oral sources of information which were always plentiful. Many active physicists enjoyed and profited from discussing their problems with him and on his side he took notes of these conversations and inserted them in the artificial memory, which by now was very bulky and elaborately cross-indexed extending to all of physics. On the other hand, he barely glanced at the journals and completely stopped reading any physics books.

He once said that Weyl's book on group theory and quantum mechanics was the last physics book he read. His old friends also noticed a contraction in the span of his interest. Even for him it was becoming impossible to keep up with all of physics. He began to limit himself to high-energy nuclear physics and to his direct research interests. He still kept in reserve his powerful resources for other branches of physics, if they became necessary, but he did not seek new ventures far afield.

Again he often spoke about what would happen with the passing years when old age would set in, but it is typical and meaningful that in 1946, at the end of the Los Alamos period, after a few minutes of reflection, in seriously estimating the work accomplished and the path of the future, he used the number 1/3, meaning that up to then he had accomplished about 1/3 of what he hoped to accomplish during his whole life. That the problem of aging was present to him, even if it did not worry him, is apparent from many conversations. In one of the last ones before he fell ill, I was telling him about some experiments on proton polarization that we had recently performed in Berkeley. He commented that they gave me the right to do nothing for about five years, without being labelled "senile", because after a piece of work of some importance one had the "right" to stop producing new research for some time before one had to be classed as a "retired" or "finished" scientist.

In his last years I also noted an extreme desire to avoid any waste of time, almost as if he had forebodings that time would be his most precious commodity. He worked and behaved as if he now had an obsession to avoid squandering time or energy, or wasting any of his possibilities. He acted as if he had a task assigned, a goal to reach.

In 1954 he fell ill in an insidious form. In the summer, with an effort of his iron will, he still went to Europe, made some mountain trips, and gave a beautiful course on pion physics at the summer school in Varenna, but by now his health conditions were clearly alarming. He returned to Chicago in September and submitted to an exploratory operation to diagnose the nature of his illness, which had remained obscure despite repeated examinations. Unfortunately, the operation could only recognize a hopeless situation. Fermi realized this immediately and accepted it with Socratic serenity. I believe that neither the members of his family, nor the friends who visited him during his illness, will ever forget the deep impression they received from his conversation at that time. He died on November 29, 1954, shortly after his 53rd birthday. He gave to science all he had and with him disappeared the last universal physicist in the tradition of the great men of the 19th century, when it was still possible for a single person to reach the highest summits, both in theory and experiment, and to dominate all fields of physics.

References

1. *Sulla dinamica di un sistema rigido di cariche elettriche in moto traslatorio.* «Nuovo Cimento», *22*, 199–207 (1921)

2. *Sull' elettrostatica di un campo gravitazionale uniforme e sul peso delle masse elettromagnetiche.* «Nuovo Cimento», *22*, 176–188 (1921)

6. *I raggi Röntgen.* «Nuovo Cimento», *24*, 133–163 (1922)

7. *Formazione di immagini coi raggi Röntgen.* «Nuovo Cimento», *25*, 63–68 (1923)

11*a*. I. *Beweis dass ein mechanisches Normalsystem im allgemeinen quasi-ergodisch ist.* «Phys. Zeits.», *24*, 261–265 (1923); II. *Über die Existenz quasi-ergodischer Systeme.* «Phys. Zeits.», *25*, 166–167 (1924)

11*b*. *Dimostrazione che in generale un sistema meccanico normale è quasi ergodico.* «Nuovo Cimento», *25*, 267–269 (1923)

*21*a*. *Berekeningen over de intensiteiten van spektraallijnen.* «Physica», *4*, 340–343 (1924)

21*b*. *Sopra l'intensità delle righe multiple.* «Rend. Lincei», *1*, 120–124 (1925)

22. *Sui principi della teoria dei quanti.* «Rend. Seminario matematico Università di Roma», *8*, 7–12 (1925)

29. *Sopra la teoria dei corpi solidi.* «Periodico di Matematiche», *5*, 264–274 (1925)

30. *Sulla quantizzazione del gas perfetto monoatomico.* «Rend. Lincei», *3*, 145–149 (1926)

34. *Problemi di chimica, nella fisica dell' atomo.* «Periodico di Matematiche», *6*, 19–26 (1926)

37. *Il principio delle adiabatiche e la nozione di forza viva nella nuova meccanica ondulatoria.* E. Fermi ed E. Persico. «Rend. Lincei», *4* (II), 452–457 (1926)

*38*a*. *Sopra una formula di calcolo delle probabilità.* «Nuovo Cimento», *3*, 313–318 (1926)

38*b*. *Un teorema di calcolo delle probabilità ed alcune sue applicazioni.* Tesi di Abilitazione della Scuola Normale Superiore. Pisa, 1922. *Inedito*

39. *Quantum Mechanics and the Magnetic Moment of Atoms.* «Nature» (London), *118*, 876 (1926) (Letter)

42. *Sul meccanismo dell'emissione nella meccanica ondulatoria.* «Rend. Lincei», *5*, 795–800 (1927)

43. *Un metodo statistico per la determinazione di alcune proprietà dell'atomo.* «Rend. Lincei», *6*, 602–607 (1927)

59. *L'interpretazione del principio di causalità nella meccanica quantistica.* «Rend. Lincei», *11*, 980–985 (1930); «Nuovo Cimento», *7*, 361–366 (1930)

*72*a*. *La physique du noyau atomique.* Congrès International d'électricité, Paris 1932, «C. R.», 1 sect., Rep. 22, 789–807.

72b. *Lo stato attuale della fisica del nucleo atomico.* «Ric. Scientifica», *3* (2), 101–113 (1932)

95. *Sopra lo spostamento per pressione delle righe elevate delle serie spettrali.* «Nuovo Cimento», *11*, 157–166 (1934)

107. *Artificial Radioactivity Produced by Neutron Bombardment.* Part II. E. AMALDI, O. D'AGOSTINO, E. FERMI, B. PONTECORVO, F. RASETTI and E. SEGRÈ. «Proc. Roy. Soc.» (London), Series A, *149*, 522–558 (1935)

110. *On the Recombination of Neutrons and Protons.* «Phys. Rev.», *48*, 570 (1935)

112. *Sull'assorbimento dei neutroni lenti.* I. E. AMALDI ed E. FERMI. «Ric. Scientifica», *6* (2), 344–347 (1935)

113. *Sull'assorbimento dei neutroni lenti.* II. E. FERMI ed E. AMALDI. «Ric. Scientifica», *6* (2), 443–447 (1935)

114. *Sull' assorbimento dei neutroni lenti.* III. E. AMALDI ed E. FERMI. «Ric. Scientifica», *7* (1), 56–59 (1936)

115. *Sul cammino libero medio dei neutroni nella paraffina.* E. AMALDI ed E. FERMI. «Ric. Scientifica», *7* (1), 223–225 (1936)

116. *Sui gruppi di neutroni lenti.* E. AMALDI ed E. FERMI. «Ric. Scientifica», *7* (1), 310–315 (1936)

117. *Sulle proprietà diffusione dei neutroni lenti.* E. AMALDI ed E. FERMI. «Ric. Scientifica», *7* (1), 393–395 (1936)

118a. *Sopra l'assorbimento e la diffusione dei neutroni lenti.* E. AMALDI ed E. FERMI. «Ric. Scientifica», *7* (1), 454–503 (1936)

118b. *On the Absorption and the Diffusion of Slow Neutrons.* E. AMALDI and E. FERMI. «Phys. Rev.», *50*, (1), 899–928 (1936)

119a. *Sul moto dei neutroni nelle sostanze idrogenate.* «Ric. Scientifica», *7* (2), 13–52 (1936)

119b. *On the Motion of Neutrons in Hydrogenous Substances.* Translation of 119a by G. Temmer. «Ric Scientica», *7* (2), 13 (1936)

Chapter Two

Fermi and the Elucidation of Matter

Frank Wilczek appeared on this planet about the time that Enrico Fermi left it. Wilczek, a distinguished theoretical physicist, with the passage of fifty years, puts into perspective the enormous contributions of Fermi to physics. This article, and others in this volume, reminds us of the sad loss of Fermi just as so many developments, many anticipated by him, were about to occur in physics.

❖ ❖ ❖

Frank Wilczek
FERMI AND THE ELUCIDATION OF MATTER

I feel privileged to pay tribute to the memory of Enrico Fermi on the occasion of the hundredth anniversary of his birth. Fermi has always been one of my heroes, as well as my scientific "great-grandfather" (in the line Fermi → Geoffrey Chew → David Gross → Frank Wilczek). It was also a joy and an inspiration, in preparing, to browse through his papers.

I was asked to write on the subject of "Fermi's contribution to modern physics." This task requires, of course, severe selection. Other contributors who had direct association with Fermi will discuss his remarkable achievements as teacher and scientific statesman, so clearly my job is to focus on his direct contributions to the scientific literature. That still leaves far too much material, if I am to do anything but catalog. Cataloging would be silly, as well as tedious, since Fermi's collected works, with important commentaries from his associates, are readily available [1, 2]. What I decided to do, instead, was to try identifying and following some unifying thread that could tie together Fermi's most important work. Though Fermi's papers are extraordinarily varied and always focused on specific problems, such a thread was not difficult to discern. Fermi was a prolific contributor to what I feel is the most important achievement of twentieth-century physics: an accurate and, for practical purposes, completely adequate theory of matter, based on extremely nonintuitive concepts but codified in precise and usable equations.

Atomism Transformed

Since the days of Galileo and Newton, the goal of physics—rarely articulated, but implicit in its practice—had been to derive dynamical equations so that, given the configuration of a material system at one time, its configuration at other times could be predicted. The description of the solar system, based on Newtonian celestial mechanics, realized this goal. This description excellently accounts for Kepler's laws of planetary motion, the tides, the precession of the equinoxes, and much else, but it gives no a priori predictions for such things as the number of planets and their moons, their relative sizes, or the dimensions of their orbits. Indeed, we now know that other stars support quite different kinds of planetary systems. Similarly, the great eighteenth- and nineteenth-century discoveries in electricity, magnetism, and optics, synthesized in Maxwell's dynamical equations of electromagnetism, provided a rich description of the behavior of given distributions of charges, currents, and electric and magnetic fields, but no explanation of why there should be specific reproducible forms of matter.

More specifically, nothing in classical physics explains the existence of elementary building blocks with definite sizes and properties. Yet one of the most basic facts about the physical world is that matter is built up from a few fundamental building blocks (e.g., electrons, quarks, photons, gluons), each occurring in vast numbers of identical copies. Were this not true, there could be no lawful chemistry, because every atom would have its own quirky properties. In nature, by contrast, we find accurate uniformity of properties, even across cosmic scales. The patterns of spectral lines emitted by atoms in the atmospheres of stars in distant galaxies match those we observe in terrestrial laboratories.

Qualitative and semiquantitative evidence for some form of atomism has been recognized for centuries [3]. Lucretius gave poetic expression to ancient atomism, and Newton endorsed it in his famous query 31, beginning, "It seems probable to me, that God in the beginning formed matter in solid, massy, hard, impenetrable, moveable particles, of such sizes and figures, and with such other properties, and in such proportions to space, as most conduced to the ends for which He formed them; and that these primitive particles being solids, are incomparably harder than any porous bodies compounded of them, even so very hard, as never to wear or break in pieces; no ordinary power being able to divide what God Himself made one in the first creation."

In the nineteenth century Dalton, by identifying the law of definite proportions, made atomism the basis of scientific chemistry. Rudolf Clausius, Maxwell, and Ludwig Boltzmann used it to construct successful quantitative theories of the behavior of gases. But in these developments the properties of atoms themselves were not derived, but assumed, and their theory was not much advanced beyond Newton's formulation. In particular, two fundamental questions went begging.

Question 1: Why is matter built from vast numbers of particles that fall into a small number of classes? And why do all particles of a given class rigorously display the same properties?

The indistinguishability of particles is so familiar and so fundamental to all of modern physical science that we could easily take it for granted. Yet it is by no means obvious. For example, it directly contradicts one of the pillars of Leibniz's metaphysics, his "principle of the identity of indiscernibles," according to which two objects cannot differ solely in number, but will always exhibit some distinguishing features. Maxwell thought the similarity of different molecules so remarkable that he devoted the last part of his *Encyclopedia Britannica* entry on atoms—well over a thousand words—to discussing it. He concluded, "the formation of a molecule is therefore an event not belonging to that order of nature in which we live . . . it must be referred to the epoch, not of the formation of the earth or the solar system . . . but of the establishment of the existing order of nature."

Question 2: Why are there different classes of particles? And why do they exist in the proportions they do?

As we have just seen, both Newton and Maxwell considered this question but thought that its answer lay beyond the horizon of physics.

By the end of the twentieth century, physics had made decisive progress on these questions. From a science of "how," it had expanded into a science of "what" that supported a much deeper appreciation of "why." A radically new model of matter had been constructed. The elementary building blocks had been inventoried; the equations for their behavior had been precisely defined. The building blocks of our transformed atomism are, paradoxically, both far more reproducible and far more fluid in their properties than classically conceived atoms. Fermi was a chief architect of this construction, contributing to the design at many levels, as I shall now discuss.

The Identical and the Indistinguishable

From the perspective of classical physics the indistinguishability of electrons (or other elementary building blocks) is both inessential and surprising. If electrons were nearly but not quite precisely identical, for example, if their masses varied over a range of a few parts per billion, then according to the laws of classical physics different specimens would behave in nearly but not quite the same way. And since the possible behavior is continuously graded, we could not preclude the possibility that future observations, attaining greater accuracy than is available today, might discover small differences among electrons. Indeed, it would seem reasonable to expect that differences would arise, since over a long lifetime each electron might wear down, or get bent, according to its individual history.

The first evidence that the similarity of like particles is quite precise, and goes deeper than mere resemblance, emerged from a simple but profound reflection by Josiah Willard Gibbs, in his work on the foundations of statistical mechanics. It is known as "Gibbs's paradox," and goes as follows:

Suppose that we have a box separated into two equal compartments A and B, both filled with equal densities of hydrogen gas at the same temperature. Suppose further that there is a shutter separating the compartments, and consider what happens if we open the shutter and allow the gas to settle into equilibrium. The molecules originally confined to A (or B) might then be located anywhere in A + B. Thus, since there appear to be many more distinct possibilities for distributing the molecules, it would seem that the entropy of the gas, which measures the number of possible microstates, will increase. On the other hand, one might have the contrary intuition, based on everyday experience, that the properties of gases in equilibrium are fully characterized by their volume, temperature, and density. If that intuition is correct, then in our thought experiment, the act of opening the shutter makes no change in the state of the gas, and so of course it generates no entropy. In fact, this result is what one finds in actual experiments.

The experimental verdict on Gibbs's paradox has profound implications.

If we could keep track of every molecule we would certainly have the extra entropy, the so-called entropy of mixing. Indeed, when gases of different types, say hydrogen and helium, are mixed, entropy is generated. Since entropy of mixing is not observed for (superficially) similar gases, there can be no method, even in principle, for telling their molecules apart. Thus, we cannot make a rigorous statement of the kind, "Molecule 1 is in A, molecule 2 is in A, . . . molecule n is in A," but only a much weaker statement, of the kind, "There are n molecules in A." In this precise sense, hydrogen molecules are not only similar, or even only identical, but beyond that, indistinguishable. Fermi motivated his concept of state-counting [4] through a different, though related, difficulty of classical statistical mechanics, namely, its failure to account for Nernst's law. This law, abstracted from empirical data, implied that the entropy of a material vanishes at the absolute zero of temperature. Like Gibbs's paradox it indicates that there are far fewer states than appear classically, and in particular no entropy of mixing. Fermi therefore proposed to generalize Pauli's exclusion principle from its spectroscopic origins to a universal principle, not only for electrons, but as a candidate to describe matter in general: "We will therefore assume in the following that, at most, one molecule with given quantum numbers can exist in our gas: as quantum numbers we must take into account not only those that determine the internal motions of the molecule but also the numbers that determine its translational motion."

It is remarkable, and perhaps characteristic, that Fermi uses the tested methods of the old quantum theory in evaluating the properties of his ideal gas. He places his molecules in an imaginary shallow harmonic well, identifies the single-particle energy levels by appeal to the Bohr-Sommerfeld quantization rules, and assigns one state of the system to every distribution of particles over these levels. (Of course, he does not fail to remark that the results should not depend on the details of this procedure.) All the standard results on the ideal Fermi-Dirac gas are then derived in a few strokes, by clever combinatorics.

Indeed, after a few introductory paragraphs, the paper becomes a series of equations interrupted only by minimal text of the general form, "now let us calculate . . ." So the following interpolation, though brief, commands attention: "At the absolute zero point, our gas molecules arrange themselves in a kind of shell-like structure which has a certain analogy to the arrangement of electrons in an atom with many electrons."

We can see in this the germ of the Thomas-Fermi model of atoms [5], which grows very directly out of Fermi's treatment of the quantum gas, and gives wonderful insight into the properties of matter [6]. Also, of course, the general scheme of fermions in a harmonic well is the starting point for the nuclear shell model—of which more below.

The successful adaptation of Fermi's ideal gas theory to describe electrons in metals, starting with Arnold Sommerfeld and Hans Bethe, and many other applications, vindicated his approach.

The Primacy of Quantum Fields 1: Free Fields

As I have emphasized, Fermi's state-counting logically requires the radical indistinguishability of the underlying particles. It does not explain that fact, however, so its success only sharpens the fundamental problem posed in our question 1. Deeper insight into this question requires considerations of another order. It comes from the synthesis of quantum mechanics and special relativity into quantum field theory.

The field concept came to dominate physics starting with the work of Faraday in the mid-nineteenth century. Its conceptual advantage over the earlier Newtonian program of physics, to formulate the fundamental laws in terms of forces among atomic particles, emerges when we take into account the circumstance, unknown to Newton (and, for that matter, Faraday), but fundamental in special relativity, that physical influences travel no faster than a finite limiting speed. For this implies that the force on a given particle at a given time cannot be inferred from the positions of other particles at that time, but must be deduced in a complicated way from their previous positions. Faraday's intuition that the fundamental laws of electromagnetism could be expressed most simply in terms of fields filling space and time was of course brilliantly vindicated by Maxwell's mathematical theory.

The concept of locality, in the crude form that one can predict the behavior of nearby objects without reference to distant ones, is basic to scientific practice. Practical experimenters—if not astrologers—confidently expect, on the basis of much successful experience, that after reasonable (generally quite modest) precautions isolating their experiments, they will obtain reproducible results.

The deep and ancient historic roots of the field and locality concepts provide no guarantee that these concepts remain relevant or valid when extrapolated far beyond their origins in experience, into the subatomic and quantum domain. This extrapolation must be judged by its fruits. Remarkably, the first consequences of relativistic quantum field theory supply the answer to our question 1, in its sharp form including Fermi's quantum state-counting.

In quantum field theory, particles are not the primary reality. Rather, as demanded by relativity and locality, it is fields that are the primary reality. According to quantum theory, the excitations of these fields come in discrete lumps. These lumps are what we recognize as particles. In this way, particles are derived from fields. Indeed, what we call particles are simply the form in which low-energy excitations of quantum fields appear. Thus, all electrons are precisely alike because all are excitations of the same underlying ur-stuff, the electron field. The same logic, of course, applies to photons or quarks, or even to composite objects such as atomic nuclei, atoms, or molecules.

Given the indistinguishability of a class of elementary particles, including complete invariance of their interactions under interchange, the general principles of quantum mechanics teach us that solutions forming any representation of the permutation symmetry group retain that property in time. But they do not constrain which representations are realized. Quantum field theory not only explains the existence of indistinguishable particles and the invariance of their interactions under interchange, but also constrains the symmetry of the solutions. There are two possibilities, bosons and fermions. For bosons, only the identity representation is physical (symmetric wave functions); for fermions, only the one-dimensional odd representation is physical (antisymmetric wave functions). One also has the spin-statistics theorem, according to which objects with integer spin are bosons, whereas objects with half odd-integer spin are fermions. Fermions, of course, obey Fermi's state-counting procedure. Examples are electrons, protons, neutrons, quarks, and other charged leptons and neutrinos.

It would not be appropriate to review here the rudiments of quantum field theory, which justify the assertions of the preceding paragraph. But a brief heuristic discussion seems in order.

In classical physics particles have definite trajectories, and there is no limit to the accuracy with which we can follow their paths. Thus, in principle, we could always keep tabs on who's who. Thus, classical physics is inconsistent with the rigorous concept of indistinguishable particles, and it comes out on the wrong side of Gibbs's paradox.

In the quantum theory of indistinguishable particles the situation is quite different. The possible positions of particles are described by waves (i.e., their wave functions). Waves can overlap and blur. Related to this, there is a limit to the precision with which their trajectories can be followed, according to Heisenberg's uncertainty principle. So when we calculate the quantum-mechanical probability amplitude for a physical process to take place, we must sum contributions from all ways in which it might have occurred. Thus, to calculate the amplitude that a state with two indistinguishable particles of a given sort—call them quants—at positions x_1, x_2 at time t_i will evolve into a state with two quants at x_3, x_4 at time t_f, we must sum contributions from all possible trajectories for the quants at intermediate times. These trajectories fall into two distinct categories. In one category, the quant initially at x_1 moves to x_3, and the quant initially at x_2 moves to x_4. In the other category, the quant initially at x_1 moves to x_4, and the quant initially at x_2 moves to x_3. Because (by hypothesis) quants are indistinguishable, the final states are the same for both categories. Therefore, according to the general principles of quantum mechanics, we must add the amplitudes for these two categories. We say there are "direct" and "exchange" contributions to the process. Similarly, if we have more than two quants, we must add contributions involving arbitrary permutations of the original particles.

Since the trajectories fall into discretely different classes, we may also consider the possibility of adding the amplitudes with relative factors. Mathematical consistency severely limits our freedom, however. We must demand that the rule for multiplying amplitudes, when we sum over states at an intermediate time, be consistent with the rule for the overall amplitude. Since the final result of double exchange is the same as no exchange at all, we must assign factors in such a way that *direct × direct = exchange × exchange*, and the only consistent possibilities are *direct/exchange* = ±1. These correspond to bosons (+) and fermions (−), respectively. The choice of sign determines how the interference term between direct and exchange contributions contributes to the square of the amplitude, that is, the probability for the overall process. This choice is vitally important for calculating the elementary interactions even of short-lived, confined, or elusive particles, such as quarks and gluons, whose equilibrium statistical mechanics is moot [7].

The Centerpiece: β-Decay

The preceding consequences of quantum field theory follow from its basic "kinematic" structure, independent of any specific dynamical equations. They justified Fermi's state-counting and rooted it in a much more comprehensive framework. But Fermi's own major contributions to quantum field theory came at the next stage, in understanding its dynamical implications.

Fermi absorbed quantum electrodynamics by working through many examples and using them in teaching. In so doing, he assimilated Paul Dirac's original, rather abstract formulation of the theory to his own more concrete way of thinking. His review article [8] is a masterpiece, instructive and refreshing to read even today. It begins,

> Dirac's theory of radiation is based on a very simple idea; instead of considering an atom and the radiation field with which it interacts as two distinct systems, he treats them as a single system whose energy is the sum of three terms: one representing the energy of the atom, a second representing the electromagnetic energy of the radiation field, and a small term representing the coupling energy of the atom and the radiation field.

and soon continues,

> A very simple example will explain these relations. Let us consider a pendulum which corresponds to the atom, and an oscillating string in the neighborhood of the pendulum which represents the radiation field. . . .
> To obtain a mechanical representation of this [interaction] term, let us tie the mass M of the pendulum to a point A of the string by means of a very thin and elastic thread.
> [I]f a period of the string is equal to the period of the pendulum, there is resonance and the amplitude of vibration of the pendulum becomes considerable after a certain time. This process corresponds to the absorption of radiation by the atom.

Everything is done from scratch, starting with harmonic oscillators. Fully elaborated examples of how the formalism reproduces concrete experimental arrangements in space-time are presented, including the Doppler effect and Lippmann fringes, in addition to the "S-matrix"-type scattering processes that dominate modern textbooks.

Ironically, in view of what was about to happen, Fermi's review article does not consider systematic quantization of the electron field. Various processes involving positrons are discussed on an ad hoc basis, essentially following Dirac's hole theory.

With hindsight, it appears obvious that James Chadwick's discovery of the neutron in early 1932 marks the transition between the ancient and the classic eras of nuclear physics. (The dominant earlier idea, consonant with an application of Occam's razor that in retrospect seems reckless, was that nuclei were made from protons and tightly bound electrons. This idea is economical of particles, of course, but it begs the question of dynamics, and it also has problems with quantum statistics—N^{14} would be 21 particles, and a fermion, whereas molecular spectroscopy shows it is a boson.) At the time, however, there was much confusion [9].

The worst difficulties which brought into question the validity of quantum mechanics and energy conservation in the nuclear domain concerned β-decay. On the face of it, the observations seemed to indicate a process $n \to p + e^-$, wherein a neutron decays into a proton plus electron. However, this would indicate boson statistics for the neutron, and it reintroduces the spectroscopic problems mentioned above. Moreover, electrons are observed to be emitted with a broad spectrum of energies, which is impossible for a two-body decay if energy and momentum are conserved. Bohr in this context suggested abandoning conservation of energy. However, Wolfgang Pauli suggested that the general principles of quantum theory could be respected and the conservation laws could be maintained if the decay were in reality $n \to p + e^- + \bar{\nu}$, with a neutral particle, $\bar{\nu}$, escaping detection.

Fermi invented the term "neutrino" for Pauli's particle. This started as a little joke in conversation, *neutrino* being the Italian diminutive of *neutrone*, suggesting the contrast of a little neutral one versus a heavy neutral one. It stuck, of course. (Actually, what appears in neutron decay is what we call today an antineutrino.)

More important, Fermi took Pauli's idea seriously and literally and attempted to bring it fully into line with special relativity and quantum mechanics. This meant constructing an appropriate quantum field theory. Having mastered the concepts and technique of quantum electrodynamics, and after absorbing the Jordan-Wigner technique for quantizing fermion fields, he was ready to construct a quantum field theory for β-decay. He chose a Hamiltonian of the simplest possible type, involving a local coupling of the four fields involved, one for each particle created or destroyed. There are various possibilities for combining the spinors into a Lorentz-invariant object, as Fermi discusses. He then calculates the electron emission spectrum, including the possible effect of nonzero neutrino mass.

It soon became apparent that these ideas successfully organize a vast wealth of data on nuclear β-decays in general. They provided, for forty years thereafter, the ultimate foundation of weak interaction theory, and still remain the most useful working description of a big chapter of nuclear and particle physics. Major refinements including the concept of universality, parity violation, and V-A theory—which together made the transition to a deeper foundation, based on the gauge principle, compelling—not to mention the experimental investigation of neutrinos themselves, were all based firmly on the foundation supplied by Fermi.

The Primacy of Quantum Fields 2:
From Local Interactions to Particles and Forces (Real and Virtual)

Although Fermi's work on β-decay was typically specific and sharply focused, by implication it set a much broader agenda for quantum field theory. It emphasized the very direct connection between the abstract principles of interacting quantum field theory and a most fundamental aspect of nature, *the ubiquity of particle creation and destruction processes.*

Local interactions involve products of field operators at a point. When the fields are expanded into creation and annihilation operators multiplying modes, we see that these interactions correspond to processes wherein particles can be created, annihilated, or changed into different kinds of particles. This possibility arises, of course, in the primeval quantum field theory, quantum electrodynamics, where the primary interaction arises from a product of the electron field, its Hermitean conjugate, and the photon field. Processes of radiation and absorption of photons by electrons (or positrons), as well as electron-positron pair creation, are encoded in this product. But because the emission and absorption of light is such a common experience, and electrodynamics such a special and familiar classical field theory, this correspondence between formalism and reality did not initially make a big impression. The first conscious exploitation of the potential for quantum field theory to describe processes of transformation was Fermi's theory of β-decay. He turned the procedure around, inferring from the observed processes of particle transformation the nature of the underlying local interaction of fields. Fermi's theory involved creation and annihilation not of photons but of atomic nuclei and electrons (as well as neutrinos)— the ingredients of "matter." It thereby initiated the process whereby classical atomism, involving stable individual objects, was replaced by a more sophisticated and accurate picture. In this picture only the fields, and not the individual objects they create and destroy, are permanent.

This line of thought gains power from its association with a second general consequence of quantum field theory, *the association of forces and interactions with particle exchange.* When Maxwell completed the equations of electrodynamics, he found that they supported source-free electromagnetic waves. The classical electric and magnetic fields took on a life of their own. Electric and magnetic forces between charged particles are explained as due to one particle acting as a source for electric and magnetic fields, which then influence others. With the correspondence of fields and particles, as it arises in quantum field theory, Maxwell's discovery corresponds to the existence of photons, and the generation of forces by intermediary fields corresponds to the exchange of virtual photons.

The association of forces (or more generally, interactions) with particles is a general feature of quantum field theory. "Real" particles are field excitations that can be considered usefully as independent entities, typically because they have a reasonably long lifetime and can exist spatially separated from other excitations, so that we can associate transport of definite units of mass, energy, charge, and so on, with them. But in quantum theory all excitations can also be produced as short-lived fluctuations. These fluctuations are constantly taking place in what we regard as empty space, so that the physicists' notion of vacuum is very far removed from simple nothingness. The behavior of real particles is affected by their interaction with these virtual fluctuations. Indeed, according to quantum field theory that's all there is! So observed forces must be ascribed to fluctuations of quantum fields—but these fields will then also support genuine excitations, that is, real particles. Tangible particles and their "virtual" cousins are as inseparable as two sides of the same coin. This connection was used by Hideki Yukawa to infer the existence and mass of pions from the range of nuclear forces. (Yukawa began his work by considering whether the exchange of virtual electrons and neutrinos, in Fermi's β-decay theory, might be responsible for the nuclear force! After showing that these virtual particles gave much too small a force, he was led to a new particle.) More recently it has been used in electroweak theory to infer the existence, mass, and properties of W and Z bosons prior to their observation, and in quantum chromodynamics (QCD) to infer the existence and properties of gluon jets prior to their observation.

This circle of ideas, which to me forms the crowning glory of twentieth-century physics, grew around Fermi's theory of β-decay. There is a double meaning in my title, "Fermi and the Elucidation of Matter." For Fermi's most beautiful insight was, precisely, to realize the profound commonality of matter and light.

Nuclear Chemistry

The other big classes of nuclear transformations, of quite a different character from β-decay, are those in which no leptons are involved. In modern language, these are processes mediated by the strong and electromagnetic interactions. They include the fragmentation of heavy nuclei (fission) and the joining of light nuclei (fusion). These processes are sometimes called nuclear chemistry, since they can be pictured as rearrangements of existing materials—protons and neutrons—similar to how ordinary chemistry can be pictured as rearrangements of electrons and nuclei. In this terminology, it would be natural to call β-decay nuclear alchemy.

Fermi discovered the technique that, above all others, opened up the experimental investigation of nuclear chemistry. This is the potent ability of slow neutrons to enter and stir up nuclear targets. Fermi regarded this as his greatest discovery. In an interview with S. Chandrasekhar, quoted in [2], he described it:

> I will tell you now how I came to make the discovery which I suppose is the most important one I have made. We were working very hard on the neutron-induced radioactivity and the results we were obtaining made no sense. One day, as I came to the laboratory, it occurred to me that I should examine the effect of placing a piece of lead before the incident neutrons. Instead of my usual custom, I took great pains to have the piece of lead precisely machined. I was clearly dissatisfied with something: I tried every excuse to postpone putting the piece of lead in its place. When finally, with some reluctance, I was going to put it in its place, I said to myself: "No, I do not want this piece of lead here; what I want is a piece of paraffin." It was just like that, with no advance warning, no conscious prior reasoning. I immediately took some odd piece of paraffin and placed it where the piece of lead was to have been.

There are wonderful accounts of how he mobilized his group in Rome to exploit, with extraordinary joy and energy, his serendipitous discovery [2]. I will not retell that story here, or the epic saga of the nuclear technology commencing with the atomic bomb project [10]. We are still coming to terms with the destructive potential of nuclear weapons, and we have barely begun to exploit the resource of nuclear energy.

From the point of view of pure physics, the significance of Fermi's work in nuclear chemistry was above all to show that in a wealth of phenomena nuclei could be usefully described as composites wherein protons and neutrons, though in close contact, retain their individual identity and properties. This viewpoint reached its apex in the shell model of Maria Mayer and Hans Jensen [11]. Famously, Fermi helped inspire Mayer's work, and in particular suggested the importance of spin-orbit coupling, which proved crucial. The success of Mayer's independent-particle model for describing quantum systems in which the interactions are very strong stimulated deep work in the many-body problem. It also provided the intellectual background for the quark model.

Last Insights and Visions

With developments in nuclear chemistry making it clear that different nuclei are built up from protons and neutrons, and developments in β-decay theory showing how protons and neutrons can transmute into one another, our question 2, to understand the specific content of the world, came into sharp focus. Specifically, it became possible to pose the origin of the elements as a scientific question. Fermi was very much alive to this possibility that his work had opened up. With Anthony Turkevich he worked extensively on George Gamow's "ylem" proposal that the elements are built up by successive neutron captures, starting with neutrons only in a hot, rapidly expanding universe. They correctly identified the major difficulty with this idea, that is, the insurmountable gap at atomic number 5, where no suitably stable nucleus exists. Yet a rich and detailed account of the origin of elements can be constructed, very nearly along these lines. The Fermi-Turkevich gap at A = 5 is reflected in the observed abundances, in that less than 1% of cosmic chemistry resides in such nuclei, the astronomers' "metals." Elements beyond A = 5 (except for a tiny amount of Li^7) are produced in a different way, in the reactions powering stars, and are injected back into circulation through winds or, perhaps, with additional last-minute cooking, during supernova explosions. Also, the correct initial condition for big-bang nucleosynthesis postulates thermal equilibrium at high temperatures, not all neutrons. Though he didn't quite get there, Fermi envisioned this promised land.

All this progress in observing, codifying, and even controlling nuclear processes was primarily based on experimental work. The models used to correlate the experimental data incorporated relativistic kinematics and basic principles of quantum mechanics, but were not derived from a closed set of underlying equations. The experimental studies reveal that the interaction between nucleons at low energy is extremely complex. It depends on distance, velocity, and spin in a completely entangled way. One could parameterize the experimental results in terms of a set of energy-dependent functions—"phase shifts"—but these functions displayed no obvious simplicity.

The lone triumph of high-energy theory was the discovery of Yukawa's pion. This particle, with the simple local coupling postulated by Yukawa's theory, could account semiquantitatively for the long-range tail of the force. Might it provide the complete answer? No one could tell for sure—the necessary calculations were too difficult.

Starting around 1950 Fermi's main focus was on the experimental investigation of pion-nucleon interactions. They might be expected to have only the square root of the difficulty in interpretation, so to speak, since they are closer to the core element of Yukawa's theory. But pion-nucleon interactions turned out to be extremely complicated also.

With the growing complexity of the observed phenomena, Fermi began to doubt the adequacy of Yukawa's theory. No one could calculate the consequences of the theory accurately, but the richness of the observed phenomena undermined the basis for hypothesizing that with pointlike protons, neutrons, and pions one had reached bottom in the understanding of strongly interacting matter. There were deeper reasons for doubt, arising from a decent respect for question 2. Discovery of the μ particle, hints of more new particles in cosmic ray events (eventually evolving into our K mesons), together with the familiar nucleons, electrons, and photons, plus neutrinos and the pions, indicated a proliferation of "elementary" particles. They all transformed into one another in complicated ways. Could one exploit the transformative aspect of quantum field theory to isolate a simple basis for this proliferation—fewer, more truly elementary building blocks?

In one of his late theoretical works, with C. N. Yang, Fermi proposed a radical alternative to Yukawa's theory that might begin to economize particles. They proposed that the pion was not fundamental and elementary at all, but rather a composite particle, specifically a nucleon-antinucleon bound state. This was a big extrapolation of the idea behind the shell model of nuclei. Further, they proposed that the primary strong interaction was what we would call today a four-fermion coupling of the nucleon fields. The pion was to be produced as a consequence of this interaction, and Yukawa's theory as an approximation—what we would call today an effective field theory. The primary interaction in the Fermi-Yang theory is the same form of interaction that appears in Fermi's theory of β-decay, though of course the strength of the interaction and the identities of the fields involved are quite different. In his account of this work [12] Yang says, "As explicitly stated in the paper, we did not really have any illusions that what we suggested may actually correspond to reality. . . . Fermi said, however, that as a student one solves problems, but as a research worker one raises questions."

Indeed, the details of their proposal do not correspond to our modern understanding. In particular, we have learned to be comfortable with a proliferation of particles, so long as their fields are related by symmetries. But some of the questions Fermi and Yang raised—or, I would say, the directions they implicitly suggested—were, in retrospect, fruitful ones. First, the whole paper is firmly set in the framework of relativistic quantum field theory. Its goal, in the spirit of quantum electrodynamics and Fermi's β-decay theory, was to explore the possibilities of that framework, rather than to overthrow it. For example, at the time the existence of antinucleons had not yet been established experimentally. However, the existence of antiparticles is a general consequence of relativistic quantum field theory, and it is accepted with minimal comment. Second, building up light particles from much heavier constituents was a liberating notion. It is an extreme extrapolation of the lowering of mass through binding energy. Nowadays we go much further along the same line, binding infinitely heavy (confined) quarks and gluons into the observed strongly interacting particles, including both pions and nucleons.

Third, and most profound, the possibility of deep similarity in the basic mechanism of the strong and the weak interaction, despite their very different superficial appearance, is anticipated. The existence of just such a common mechanism, rooted in concepts later discovered by Fermi's coauthor—Yang-Mills theory—is a central feature of the modern theory of matter, the so-called standard model.

In another of his last works, with John Pasta and Stanisław Ulam [13], Fermi enthusiastically seized upon a new opportunity for exploration—the emerging capabilities of rapid machine computation. With his instinct for the border of the unknown and the accessible, he chose to revisit a classic, fundamental problem that had been the subject of one of his earliest papers: the problem of approach to equilibrium in a many-body system. The normal working hypothesis of statistical mechanics is that equilibrium is the default option and is rapidly attained in any complex system unless some simple conservation law forbids it. But proof is notoriously elusive, and Fermi thought to explore the situation by controlled numerical experiments, wherein the degree of complexity could be varied. Specifically, he considered various modest numbers of locally coupled nonlinear springs. A stunning surprise emerged: The approach to equilibrium is far from trivial; there are emergent structures, collective excitations that can persist indefinitely. The topic of solitons, which subsequently proved vast and fruitful, was foreshadowed in this work. And the profound but somewhat nebulous question, central to an adequate understanding of nature, of how ordered structures emerge spontaneously from simple homogeneous laws and minimally structured initial conditions, itself emerged spontaneously.

Not coincidentally, the same mathematical structure—locally coupled nonlinear oscillators—lies at the foundation of relativistic quantum field theory. Indeed, as we saw, this was the way Fermi approached the subject right from the start. In modern QCD the emergent structures are protons, pions, and other hadrons, which are well hidden in the quark and gluon field "springs." Numerical work of the kind Fermi pioneered remains our most reliable tool for studying these structures.

Clearly, Fermi was leading physics toward fertile new directions. When his life was cut short, it was a tremendous loss for our subject.

Fermi as Inspiration: Passion and Style

Surveying Fermi's output as a whole, one senses a special passion and style, unique in modern physics. Clearly, Fermi loved his dialogue with nature. He might respond to her deepest puzzles with queries of his own, as in the explorations we have just discussed, or with patient gathering of facts, as in his almost brutally systematic experimental investigations of nuclear transmutations and pion physics. But he also took great joy in solving, or simply explicating, her simpler riddles, as the many Fermi stories collected in this volume attest.

Fermi tackled ambitious problems at the frontier of knowledge, but always with realism and modesty. These aspects of his scientific style shine through one of Fermi's rare "methodological" reflections, written near the end of his life [14]:

> When the Yukawa theory first was proposed there was a legitimate hope that the particles involved, protons, neutrons, and pi-mesons, could be legitimately considered as elementary particles. This hope loses more and more its foundation as new elementary particles are rapidly being discovered.
>
> It is difficult to say what will be the future path. One can go back to the books on method (I doubt whether many physicists actually do this) where it will be learned that one must take experimental data, collect experimental data, organize experimental data, begin to make working hypotheses, try to correlate, and so on, until eventually a pattern springs to life and one has only to pick out the results. Perhaps the traditional scientific method of the textbooks may be the best guide, in the lack of anything better. . . .
>
> Of course, it may be that someone will come up soon with a solution to the problem of the meson, and that experimental results will confirm so many detailed features of the theory that it will be clear to everybody that it is the correct one. Such things have happened in the past. They may happen again. However, I do not believe that we can count on it, and I believe that we must be prepared for a long, hard pull.

Those of you familiar with the subsequent history of the strong interaction problem will recognize that Fermi's prognosis was uncannily accurate. A long period of experimental exploration and pattern recognition provided the preconditions for a great intellectual leap and synthesis. The process would, I think, have gratified Fermi but not surprised him. The current situation in physics is quite different from what Fermi lived through or, at the end, described. After the triumphs of the twentieth century, it is easy to be ambitious. Ultimate questions about the closure of fundamental dynamical laws and the origin of the observed universe begin to seem accessible. The potential of quantum engineering, and the challenge of understanding how one might orchestrate known fundamentals into complex systems, including powerful minds, beckon. Less easy, perhaps, is to remain realistic and (therefore) appropriately modest—in other words, to craft important subquestions that we can answer definitively, in dialogue with nature. In this art Fermi was a natural grand master, and the worthy heir of Galileo.

Acknowledgments

This work was supported in part by funds provided by the U.S. Department of Energy (D.O.E.) under cooperative research agreement DF-FC02-94ER40818. I would like to thank M. Stock for help with the manuscript.

References

1. E. Fermi, *Collected Works*, 2 vols. (University of Chicago Press, 1962–65).

2. Another indispensable source is the scientific biography *Enrico Fermi, Physicist*, by E. Segrè (University of Chicago Press, 1970).

3. An accessible selection of original sources and useful commentaries is *The World of the Atom*, ed. H. Boorse and L. Motz, 2 vols. (Basic Books, 1966).

4. Papers 30 and 31 in [1]. A portion is translated into English in [3].

5. Papers 43–48 in [1].

6. E. Lieb, "Thomas-Fermi and Related Theories of Atoms and Molecules," *Reviews of Modern Physics* 53, no. 4, pt. 1 (1981): 603–641.

7. Actually, the statistical mechanics of quarks and gluons may become accessible in ongoing experiments at the Relativistic Heavy Ion Collider at the Brookhaven National Laboratory.

8. Paper 67 in [1].

9. A. Pais, *Inward Bound: Of Matter and Forces in the Physical World* (Oxford University Press, 1986).

10. R. Rhodes, *The Making of the Atomic Bomb* (Simon and Schuster, 1986).

11. M. Mayer, *Elementary Theory of Nuclear Shell Structure* (Wiley, 1955).

12. C. N. Yang, introduction to paper 239 in [1].

13. Paper 266 in [1].

14. Paper 247 in [1].

Chapter Three

Letters and Documents Relating to the Development of Nuclear Energy

The story of the development of the chain reaction has been well documented. We include in this chapter some letters and documents which are perhaps less familiar and illustrate Fermi's personal involvement and his down-to-earth nature.

News of the sensational discovery of the fission of uranium by Hahn and Strassman was brought to the United States by Niels Bohr in early 1939. Fermi, who had just arrived in New York on January 2, met Bohr at the New York pier on Jan 16. Bohr arrived in the company of his son Erik and the physicist Leon Rosenfeld.

Bohr did not reveal the discovery to Fermi, as he was awaiting news of the direct detection of fission fragments in an experiment being carried out by Otto Frisch and Lise Meitner at Bohr's institute in Denmark. Bohr was concerned for the priority of the experiment. A few days later in a seminar at Princeton, Rosenfeld spilled the beans, and the news spread quickly. By January 25, Fermi and his colleagues at Columbia detected uranium fission in their own experiment. (There are several versions of the arrival of the news of the fission discovery. I follow the story as described in *Niels Bohr's Times: In Physics, Philosophy, and Polity,* by Abraham Pais [Oxford, 1991], chap. 20, p. 455.)

FERMI TO HARRY M. DURNING, JANUARY 16, 1939

Fermi had barely arrived in the United States when he arranged to meet Bohr at the dock. The following letter, to a New York customs officer, requests two separate pier tickets, as he and Laura would arrive to meet Bohr from separate locations. He has Laura do the "leg work."

January 16, 1939

Mr. Harry M. Durning,
Assistant Collector,
U.S. Office of Collector of Customs,
New York City.

Dear Sir:

I am writing to ask if you will kindly issue a pier permit to me and also one to Mrs. Fermi to meet Professor and Mrs. Nils Bohr who are to arrive on the Drottingsholm today. Professor and Mrs. Bohr of Copenhagen are old friends of ours whom we recently visited in Copenhagen and we wish to meet them on their arrival here.

This letter will be handed to you by Mrs. Fermi. Separate pier tickets are requested because we shall not be able to go down together, and Mrs. Fermi will have to send one of the tickets uptown to me.

Very truly yours,

EF:H

Enrico Fermi
Professor of Physics
Columbia University.

GEORGE B. PEGRAM TO S. C. HOOPER, MARCH 18, 1939

A few months later George Pegram, dean and professor of physics at Columbia University, thought that the work of Fermi could possibly lead to explosives that would be one million times more powerful than chemical explosives. Pegram wrote a letter to Admiral Hooper in the Office of Naval Operations, outlining the importance of Fermi's work and introducing En-

Dear Fermi —
This may prepare the
way for you a little better
than Mr. Compton's explanation
to Adm. Hooper.

March 18, 1939

Admiral S. C. Hooper,
Office of Chief of Naval Operations,
Navy Department,
Washington, D. C.

Dear Sir:

This morning I had a telephone conversation with Mr. Compton in the office of the Assistant Secretary of the Navy, who has doubtless reported the conversation to you. It had to do with the possibility that experiments in the physics laboratories at Columbia University reveal that conditions may be found under which the chemical element uranium may be able to liberate its large excess of atomic energy, and that this might mean the possibility that uranium might be used as an explosive that would liberate a million times as much energy per pound as any known explosive. My own feeling is that the probabilities are against this, but my colleagues and I think that the bare possibility should not be disregarded, and I therefore telephoned to Mr. Edison's office this morning chiefly to arrange a channel through which the results of our experiments might, if the occasion should arise, be transmitted to the proper authorities in the United States Navy.

Professor Enrico Fermi who, together with Dr. Szilard, Dr. Zinn, Mr. Anderson and others, has been working on this problem in our laboratories, went to Washington this afternoon to lecture before the Philosophical Society in Washington this evening and will be in Washington tomorrow. He will telephone your office, and if you wish to see him will be glad to tell you more definitely what the state of knowledge on this subject is at present.

rico Fermi. This led to a presentation to the navy by Fermi on March 18, 1939. As a result a small grant of $1500 was given to Columbia to support the work. This was the first government support that ultimately led to the $2 billion Manhattan Project. There is a handwritten note to Fermi from Pegram explaining the purpose of the letter. The name Mr. Compton is not to be confused with Arthur Compton, the physicist. Nor indeed is Mr. Edison to be confused with the famous inventor!

Admiral S. C. Hooper, page 2   March 16, 1939

            [Professor Fermi, formerly of Rome, is Professor
of Physics at Columbia University. In December last he
was awarded the Nobel Prize in Physics for 1938 for the
work that he did on the artificial creation of radioactive
elements by means of neutrons.] There is no man more
competent in this field of nuclear physics than Professor
Fermi.

            Professor Fermi has recently arrived to stay
permanently in this country and will become an American
citizen in due course. [He is very much at home in this
country, having visited here often to lecture at the
University of Michigan, Stanford University and at
Columbia.

            Professor Fermi will be staying tomorrow with]
Professor Edward Teller of George Washington University,

            Sincerely yours,

GBP:H   George B. Pegram
                            Professor of Physics

BETWEEN FERMI AND LEO SZILARD, JULY 3–11, 1939

There was an exchange of letters between Leo Szilard and Fermi in the summer of 1939. Szilard was in New York and Fermi was attending the Summer School of Theoretical Physics at Ann Arbor. Szilard produced a flurry of letters dated July 3, July 5, and July 8, 1939, all concerning the use of carbon (graphite) as the moderator for the production of a chain reaction. Fermi's typewritten response on July 9 was to Szilard's letter of July 3, with a handwritten note acknowledging Szilard' letter July 5. Szilard's reply to Fermi was dated July 11.

Fermi was concerned, as was Szilard, about the thermal neutron capture cross section of carbon. Fermi questions Szilard's proposed technique to measure the absorption. Fermi and Herbert Anderson ultimately measured the cross section to be 0.003 barns, which was sufficiently small to allow graphite as a moderator. There is only one experiment where Fermi and Szilard collaborated (*Physical Review* 56 [1939]: 284). It was a measurement of neutron production and absorption in uranium showing that there was an excess in production over absorption. Szilard refers to that paper in his letter of July 11. Very shortly after this exchange of letters Szilard and Eugene Wigner visited Albert Einstein, and the famous letter of Albert Einstein to President Roosevelt, dated August 2, 1939, was produced. This letter was the genesis of what became the Manhattan Project. It was also Szilard's effort which, in early 1940, led to the shipment of 1.5 tons of graphite to Columbia. Fermi and Anderson in 1940 did the experiment that showed that graphite was a practical moderator.

More about Fermi and Szilard appears in a contribution by Nina Byers in chapter 8 of this volume.

Hotel King's Crown
420 West 116th Street
New York City

July 3rd, 1939

Dear Fermi:

This is to keep you informed of the trend of my ideas concerning chain reactions. It seems to me now that there is a good chance that carbon might be an excellent element to use in place of hydrogen, and there is a strong temptation to gamble on this chance. The capture cross-section of carbon is not known: the only experimental evidence available asserts an upper limit of 0.01 times $10^{-24} cm^2$. If the cross-section were 0.01 carbon would be no better than hydrogen, but the cross-section is perhaps much smaller, and it might be for instance 0.001. If it were so carbon not only could be used in place of hydrogen, but would have great advantages, even if a chain reaction were possible with hydrogen also. The concentration of uranium oxide in carbon could be kept very low, so that one could have about 2 gm of carbon per cc. This compares favorably with 1/2 gm of water per cc at the most and means that the mean square of the displacement of a neutron for slowing down to thermal velocities would be only 1.5 times as large in the carbon-uranium-oxide mixture than in the water-uranium-oxide mixture. If capture by carbon can be neglected, the concentration of uranium oxide is determined by the consideration that the average displacement

-2-

of a thermal neutron for capture by uranium in the mixture must
not become too large. With this as a limiting factor about 1/10
of the weight of the mixture would have to be uranium, and that
means that one would need only a few tons of uranium oxide if
our present data about uranium are correct.

I personally would be in favor of trying a large scale
experiment with a carbon-uranium-oxide mixture if we can get
hold of the material.

I intend to plunge in the meantime into an experiment designed for measuring small capture cross-sections for thermal
neutrons. This is the proposed experiment: A sphere of carbon
of 20 cm radius or larger is surrounded by water and a neutron
source is placed in the center of the sphere. The slow neutron
density is measured inside the carbon sphere by an indium or
rhodium indicator at two points, one close to the surface, and
one close to the center. The slow neutron density at these two
points is measured once with, and once without, an absorbing
layer of boron (or cadmium), covering the surface of the sphere.
It is easy to calculate from the observed ratio of the differences (of the observed neutron density with and without absorber
at the surface of the sphere) obtained for the two points and
the scattering cross-section the ratio of the capture cross-
section to the scattering cross-section for thermal neutrons.
I calculate that a ratio of the neutron densities of the order
of magnitude of 75 to 100 would for instance be obtained for
two points in a sphere of carbon of about 20 cm radius if the
capture cross-section of carbon were 0.005. It seems that very

-3-

small capture cross-sections can conveniently be measured by
this method.

If carbon should fail, our next best guess might be heavy
water, and I have therefore taken steps to find out if it is
physically possible to obtain a few tons of heavy water. Heavy
hydrogen is supposed to have a capture cross-section below 0.003,
and the scattering cross-section ought to be 3 or 4 times 10^{-24}
for neutrons above the 1 volt region. (It is 6 to 7 times 10^{-24}
for the thermal region). Since heavy hydrogen slows down about
as efficiently per collision as ordinary hydrogen, and since
hydrogen has a capture cross-section of 0.27 and a scattering
cross-section of 20, heavy hydrogen is more favorable.

Yours,

(Leo Szilard)

Hotel King's Crown
420 West 116th Street
New York City

July 5th, 1939

Dear Fermi:

I think the letter I wrote you on July 3rd contains a mistake insofar as the ratio of the thermal neutron density at the center of the sphere and at the surface of the sphere is not 75 to 100, but 95 to 100 for the values given in that letter. The thermal neutron density within the sphere obeys the equation

$$D \frac{d^2(r\rho)}{dr^2} - A(r\rho) = 0$$

with

$$D = \frac{w \lambda_{sc}}{3} \; ; \; A = \frac{w}{\sigma_c} \lambda_s$$

it is:

$$\rho(r) = C \frac{e^{ar} - e^{-ar}}{r}$$

where

$$a = \frac{\sqrt{3}}{\lambda_{sc}} \sqrt{\frac{\sigma_c}{\sigma_{sc}}}$$

For small ar we have $\rho(r) = 2aC \left(1 + \frac{a^2 r^2}{6}\right)$ and the ratio of the densities on the surface and in the center is given by $\frac{\rho(r)}{\rho(0)} = \frac{e^{ar} - e^{-ar}}{2ar} \simeq 1 + \frac{a^2 r^2}{6}$.

For $r = 20$ cm, $\lambda_{sc} = 2$ cm, and $\frac{\sigma_c}{\sigma_{sc}} \simeq \frac{1}{1000}$ we have $\frac{\rho(r)}{\rho(0)} \simeq 1.05$.

As you see, the method is beginning to get somewhat awkward in the case of carbon for smaller capture cross-sections than 0.005. It seems that it will be possible to get sufficiently pure carbon at a reasonable price. Carbon would also have an advantage over

-2-

hydrogen insofar as there is no change in the
scattering cross-section in the transition from
the resonance region to the thermal region. Consequently, if layers of uranium oxide of finite
thickness are used, the diffusion of the thermal
neutrons produced in the carbon to the uranium
layer is not adversely affected as in the case
of hydrogen by such a change. Whether this point
is of any importance depends of course on the
absolute value of the carbon cross-section. Pend-
ing reliable information about carbon we ought
perhaps to consider heavy water as the "favorite",
and I shall let you know as soon as I can how
many tons could be obtained within reasonable time.

With kind regards to all,

Yours,

(Leo Szilard)

④

Hotel King's Crown
420 West 116th Street
New York City

July 8th, 1939

Dear Fermi:

Sorry to bombard you with so many letters about carbon.
This is just to tell you that I have reached the conclusion
that it would be the wisest policy to start a large scale experiment with carbon right away without waiting for the outcome
of the absorption measurement which was discussed in my last
two letters, The two experiments might be done simultaneously.
The following can be said in favor of this procedure:

A chain reaction with carbon is so much more convenient
and so much more important from the point of view of applications than a chain reaction with heavy water or helium that
we must know in the shortest possible time whether we can make
it go. This can be decided with certainty in a relatively short
time by a large scale experiment, and therefore this experiment
ought to be performed. If we waited for the absorption measurement we would lose three months, and in case the result is positive we would still not know with a 100% certainty the answer
with respect to the question of the chain reaction.

I thought that perhaps 50 tons of carbon and 600/tons of
uranium should be used as a start. The value of the carbon would
only be about $ 10.000.- Since the carbon and the uranium oxide

-2-

would not be mixed but built up in layers, or in any case used
in some canned form, there will be no waste of material or waste
of labor involved in unmixing after the experiment is over.
Since the uranium layers may be separated by carbon layers of
20 to 30 cm thickness, or even more, we have to deal with a
comparatively simple structure. Much simpler than would be the
case for alternating water and uranium layers.

I told Professor Pegram yesterday how I felt about the
situation, and he seemed to be not unwilling to take the necessary action. I wonder whether you think it wise to proceed
as outlined in this letter.

With kindest regards,

yours,

(Leo Szilard)

UNIVERSITY OF MICHIGAN
ANN ARBOR

DEPARTMENT OF PHYSICS   July 9 1939

Dear Szilard,
              Thank you for your letter. I was also considering the
possibility of using carbon fo slowing down the neutrons; in the obviously optimistic hypothesis that carbon should have no absorption at all
for neutrons, and assuming for the resonance absorption band of uranium
the usual data (which also I rather suspect to be optimistic) one
finds from an elementary calculation that the ratio of the concentrations
(ratio of the numbers of atoms) of uranium and carbon should be about
1½ one thousandth in order to avoid too much resonance absorption.
According to my estimates a possible recipe might be about 30000 Kg.
of carbon mixed with 600 Kg of uranium. If it were really so the amounts
of materials would certainly not be too large.
              Since however the amount of uranium that can be used, especially
in a homogeneous mixture is exceedingly small, even a very small absorption
by carbon either at thermal energy or even before might be sufficient
for preventing the chain reaction; perhaps the use of thick layers of
carbon separated by layers of uranium might allow to use a somewhat
larger percentage of uranium.
              I have been thinking about the experiment that you propose for
measuring the small absorption cross section in carbon. It seems to me
that you have probably over estimated the difference between rand and
center activity in the carbon sphere; moreover I dont see how you can
take into account the contribution of those neutrons that become thermal
due to impacts against carbon. Their number should probably not be very
large, but might disturb very considerably the measurement of a small difference.
              I had discarded heavy water as too expensive; but if you can
easily get several tons of it it might work very nicely.
              The cyclotron here will start working again next week and
I hope to be able to get reliable information on the so called resonance
absorption of uranium. I shall inform you of the results.

                              Yours sincerely

                              Enrico Fermi

P.S. I have received your second letter. If heavy
water is too expensive, as I believe, it would be
important to find some way of knowing som

thing of the carbon absorption. It seems to
me that the use of very thick layers of
C might do the trick.

Yours

Enrico Fermi

④

Hotel King's Crown
420 West 116th Street
New York City

July 11th, 1939

Dear Fermi:

Many thanks for your letter of July 9th. It obviously crossed with my third letter about carbon which probably reached you on Monday. Today, being in a hurry, I confine myself to discuss one point which you mentioned. You write with reference to the carbon sphere experiment that it might be difficult to take into account the distribution of those neutrons which become thermal due to impacts against carbon, and I wish to say the following in this connection.

The number of such neutrons which become thermal within the carbon is quite large, but their number is taken fully into account by the proposed method.

The density of the thermal neutrons within the carbon obeys the equation

$$D \frac{d^2 \rho}{dr^2} - A(r\rho) + f(r) = 0$$

where $f(r)$ stands for the number of thermal neutrons produced in unit time and unit volume at any point within the carbon sphere of the radius.

Let $\rho_1(r)$ be a solution of this equation for the boundary condition $\rho_1(R) = b$ where b is the thermal neutron density at the boundary surface of the carbon sphere in water under the conditions of the experiment.

⑤

-2-

Let further be $\rho_2(r)$ a solution of the same equation for the boundary condition $\rho_2(R) = 0$ which is realized by covering the surface of the carbon sphere with a thermal neutron absorber. The equation

$$D \frac{d^2(r\rho)}{dr^2} - A(r\rho) = 0$$

will then be obeyed by $\rho = \rho_1 - \rho_2$, and ρ will satisfy the boundary condition $\rho(R) = b$.   Therefore we have

$$\frac{\rho(r)}{\rho(0)} = \frac{\rho_1(r) - \rho_2(r)}{\rho_1(0) - \rho_2(0)} = \frac{e^{ar} - e^{-ar}}{2ar}.$$

So much for the "theory". Practical difficulties are of course present.

I may write you again in the next few days and wish to-day only to add this: Since Anderson did not get an acknowledgment from Physical Review about our note I asked Pegram today to enquire about it. It turned out that the note was too long for a Letter to the Editor and that it will appear as a short paper in the issue of August 1st.

Yours,

Y. L.

(Leo Szilard)

VANNEVAR BUSH TO FERMI, AUGUST 15, 1941

The organization for the project to develop nuclear explosives developed very slowly and did not move seriously until the Pearl Harbor attack on December 7, 1941. However in June 1941 President Roosevelt signed an executive order establishing the Office of Scientific Research and Development

OFFICE FOR EMERGENCY MANAGEMENT
OFFICE OF SCIENTIFIC RESEARCH AND DEVELOPMENT
1530 P STREET NW.
WASHINGTON, D.C.

VANNEVAR BUSH
Director

August 19, 1941

Dr. Enrico Fermi
Columbia University
New York, New York

Dear Dr. Fermi:

Upon the recommendation of Dr. James B. Conant, Chairman of the National Defense Research Committee, I have pleasure in appointing you Chairman of the subsection of Consultants on Theoretical Aspects and a member of the subsection of Consultants on Power Production, of the Uranium Section of which Dr. Briggs is chairman. The functions of the National Defense Research Committee and its relation to the Office of Scientific Research and Development are set out in the Executive Order of June 28, 1941, a copy of which is enclosed.

I hope that you will agree to undertake this highly important national service on the same basis as the civilian members of the committee, namely, on a part-time volunteer basis.

As the work on which the Committee is engaged is of a highly confidential nature, I should like to impress upon you the need for the utmost secrecy in regard to all the activities which will come to your attention in your capacity as Chairman and Consultant. The problems with which we are concerned originate in the Army and the Navy and only the high officers of these services are in a position to decide to whom the results obtained should be communicated even within the services themselves. Needless to say, the investigations cannot be discussed with any persons, civilian, military or naval except as designated by the Committee, by duly authorized Committee representatives, or by this Office.

All the members of the National Defense Research Committee have taken the usual oath of allegiance to the United States, and we are asking all people who are engaged in emergency work as part of the Committee's organization to do likewise. May I ask you, therefore, to take the enclosed oath before a justice of the peace or notary public and return the same duly witnessed to me, together with a written acknowledgment of this letter?

(OSRD) with Vannevar Bush as its director. The previously established National Defense Research Committee, with James Conant as chairman, was placed under the OSRD. Quickly Fermi was asked to be a consultant on the Uranium Section.

-2-

May I assure you of the Committee's deep appreciation, as well as my own, of your willingness to assist in these vital matters of national defense. The fact that so many scientific workers have indicated their willingness to put their immediate problems to one side and sacrifice their own personal interests is indeed heartening. I wish you success and satisfaction in service to your country.

Very truly yours,

V. Bush, Director

Enclosures

HARRY S. TRUMAN TO FERMI, AUGUST 11, 1950

President Harry Truman wrote Fermi a personal note of thanks for his service on the General Advisory Committee to the Atomic Energy Commission.

THE WHITE HOUSE
WASHINGTON

August 11, 1950

Dear Dr. Fermi:

It was a special pleasure for me to appoint to the original General Advisory Committee to the Atomic Energy Commission a man like yourself who had so much to do with the wartime development of atomic energy, and I note with real regret the expiration of your term on that Committee.

I know how often the Commission has turned to the Committee, both as a group and as individuals, and the Committee's never-failing response. I personally am deeply grateful for the thoughtful, thorough manner with which the Committee has approached its responsibilities.

Should there be need in the future, I know you will respond as generously as you have in the past to calls upon your time and effort.

Very sincerely yours,

Harry Truman

Dr. Enrico Fermi,
Professor of Physics,
Institute of Nuclear Studies,
University of Chicago,
Chicago 37,
Illinois.

OUTLINE FOR THE SPEECH "THE GENESIS OF THE NUCLEAR ENERGY PROJECT," JANUARY 30, 1954

An outline of a speech Fermi gave on the two hundredth anniversary of Columbia University, Jan 30, 1954. This is one of the few speeches that were recorded. After Fermi's death it was published in *Physics Today,* and the published speech is also reproduced here. Fermi's technique was to personally type an outline of his talk and use it as a guide for an informal delivery. It is interesting to note how closely his recorded speech follows his outline. The speech contains a great deal of humor, and one senses that Fermi was very much at ease with his audience. The speech was given on the day following Fermi's famous address as the retiring president of the American Physical Society. That speech, entitled "What Can We Learn with High Energy Accelerators," is the subject of the final chapter in this book.

The Genesis of the Nuclear Energy Project

Columbia Bicentennial - January 30   1954

Mr. Chairman....I had the opportunity of working..Pupin Lab...in the period from 1939 to 1942....Those were the years...first realized... ...first experimentation....Columbia took leading part in this phase...

I still remember...January 1939...Bohr...Fission...Hahn, Meitner, Frisch... Carnegie....Fission pulses everywhere....are neutrons emitted?

Zinn and Szilard - Anderson and myself - Joliot and von Halban - State of the Wordl and attempta at voluntary secrecy.

U 235 - Bohr and Wheeler - Nier and Dunning - Isotope separation and economy of neutrons.

Slow neutrons - Water or graphite - Not water - Liitle known of graphite.

Einstein letter - Briggs and $ 6000 - Brick laying - Ra Be sources.- Princeton (Creutz, Wigner).

Too many factors eta ef pee - Exponential- 0.87 - Purity of materials.

Pearl Harbor - Partition of responsibility - Urey and the isotope separation project- Met. Lab. at Chicago and Chain Reaction. Compton.

From then on,with some notable exceptions, the work at Columbia was concentrated on the isotope separation phase of the atomic energy project. Work initiated in 1940 by Dunning and Booth at Columbia....rapidly expanded...huge research laboratory....Carbon Carbide....Oak Ridge gas diffusion plants....this was one of the three horses on which bets....and which arrived almost simultaneously

❖ ❖ ❖

Enrico Fermi

THE GENESIS OF THE NUCLEAR ENERGY PROJECT

Mr. Chairman, Dean Pegram, fellow Members, Ladies and Gentlemen:

It seems fitting to remember, on this 200th anniversary of Columbia University, the key role that the University played in the early experimentation and the organization of the early work that led to the development of atomic energy.

I had the good fortune to be associated with the Pupin Laboratories through the period of time when at least the first phase of this development took place. I had had some difficulties in Italy and I will always be very grateful to Columbia University for having offered me a position in the Department of Physics at the most opportune moment. And in addition this offer gave me, as I said, the rare opportunity of witnessing the series of events to which I have referred.

In fact I remember very vividly the first month, January, 1939, that I started working at the Pupin Laboratories because things began happening very fast. In that period, Niels Bohr was on a lecture engagement in Princeton and I remember one afternoon Willis Lamb came back very excited and said that Bohr had leaked out great news. The great news that had leaked out was the discovery of fission and at least an outline of its interpretation; the discovery as you well remember goes back to the work of Hahn and Strassmann and at least the first idea for the interpretation came through the work of Lise Meitner and Frisch who were at that time in Sweden.

Then, somewhat later that same month, there was a meeting in Washington organized by the Carnegie Institution in conjunction with George Washington University where I took part with a number of people from Columbia University and where the possible importance of the new-discovered phenomenon of fission was first discussed in semi-jocular earnest as a possible source of nuclear power. Because it was conjectured, if there is fission with a very serious upset of the nuclear structure, it is not improbable that some neutrons will be evaporated. And if some neutrons are evaporated, then they might be more than one; let's say, for the sake of argument, two. And if they are more than one, it may be that the two of them, for example, may each one cause a fission and from that one sees of course a beginning of the chain reaction machinery.

First published in *Physics Today* 8 (November 1955): 12–16. This is a transcript of Fermi's speech at the session "Physics at Columbia University" of the American Physical Society's 1954 annual meeting.

So that was one of the things that was discussed at that conference and started a small ripple of excitement about the possibility of releasing nuclear energy. At the same time experimentation was started feverishly in many laboratories, including Pupin, and I remember before leaving Washington I had a telegram from Dunning announcing the success of an experiment directed to the discovery of the fission fragments. The same experiment apparently was at the same time carried out in half a dozen places in this country and in three or four, in fact I think slightly before, in three or four places in Europe.

Now a rather long and laborious work was started at Columbia University in order to firm up these vague suggestions that had been made as to the possibilities that neutrons were emitted and try to see whether neutrons were in fact emitted when fission took place and if so how many they would be, because clearly a matter of numbers is in this case extremely important because a little bit greater or a little bit lesser probability might have made all the difference between possibility and impossibility of a chain reaction.

Now this work was carried on at Columbia simultaneously by Zinn and Szilard on one hand and by Anderson and myself on the other hand. We worked independently and with different methods, but of course we kept close contact and we kept each other informed of the results. At the same time the same work was being carried out in France by a group headed by Joliot and von Halban. And all the three groups arrived at the same conclusion—I believe Joliot may be a few weeks earlier than we did at Columbia—namely that neutrons are emitted and they were rather abundant, although the quantitative measurement was still very uncertain and not too reliable.

A curious circumstance related to this phase of the work was that here for the first time secrecy that has been plaguing us for a number of years started and, contrary to perhaps what is the most common belief about secrecy, secrecy was not started by generals, was not started by security officers, but was started by physicists. And the man who is mostly responsible for this certainly extremely novel idea for physicists was Szilard.

I don't know how many of you know Szilard; no doubt very many of you do. He is certainly a very peculiar man, extremely intelligent (*laughter*). I see that is an understatement. (*laughter*). He is extremely brilliant and he seems somewhat to enjoy, at least that is the impression that he gives to me, he seems to enjoy startling people.

So he proceeded to startle physicists by proposing to them that given the circumstances of the period—you see it was early 1939 and war was very much in the air—given the circumstances of that period, given the danger that atomic energy and possibly atomic weapons could become the chief tool for the Nazis to enslave the world, it was the duty of the physicists to depart from what had been the tradition of publishing significant results as soon as the "Physical Review" or other scientific journals might turn them out, and that instead one had to go easy, keep back some results until it was clear whether these results were potentially dangerous or potentially helpful to our side.

So Szilard talked to a number of people and convinced them that they had to join some sort of—I don't know whether it would be called a secret society, or what it would be called. Anyway to get together and circulate this information privately among a rather restricted group and not to publish it immediately. He sent in this vein a number of cables to Joliot in France, but he did not get a favorable response from him and Joliot published his results more or less like results in physics had been published until that day. So that the fact that neutrons are emitted in fission in some abundance—the order of magnitude of one or two or three—became a matter of general knowledge. And, of course, that made the possibility of a chain reaction appear to most physicists as a vastly more real possibility than it had until that time.

Another important phase of the work that took place at Columbia University is connected with the suggestion on purely theoretical arguments, by Bohr and Wheeler, that of the two isotopes of uranium it was not the most abundant uranium 238 but it was the least abundant uranium 235, present as you know in the natural uranium mixture to the tune of 0.7 of a percent, that was responsible at least for most of the thermal fission. The argument had to do with an even number of neutrons in uranium 238 and an odd number of neutrons in uranium 235 which, according to a discussion of the binding energies that was carried out by Bohr and Wheeler, made plausible that uranium 235 should be more fissionable.

Now it clearly was very important to know the facts also experimentally and work was started in conjunction by Dunning and Booth at Columbia University and by Nier. Nier took the mass spectrographic part of this work, attempting to separate a minute but as large as possible amount of uranium 235, and Dunning and Booth at Columbia took over the part of using this minute amount in order to test whether or not it would undergo fission with a much greater cross section than ordinary uranium.

Well, you know of course by now that this experiment confirmed the theoretical suggestion of Bohr and Wheeler, indicating that the key isotope of uranium, from the point of view of any attempt of—for example—constructing a machine that would develop nuclear energy, was in fact uranium 235. Now you see the matter is important primarily for the following reasons that at the time were appreciated perhaps less definitely than at the present moment.

The fundamental point in fabricating a chain reacting machine is of course to see to it that each fission produces a certain number of neutrons and some of these neutrons will again produce fission. If an original fission causes more than one subsequent fission then of course the reaction goes. If an original fission causes less than one subsequent fission then the reaction does not go.

Now, if you take the isolated pure isotope U^{235}, you may expect that the unavoidable losses of neutrons will be minor, and therefore if in the fission somewhat more than one neutron is emitted then it will be merely a matter of piling up enough uranium 235 to obtain a chain reacting structure. But if to each gram of uranium 235 you add some 140 grams of uranium 238 that come naturally with it, then the competition will be greater, because there will be all this ballast ready to snatch away the not too abundant neutrons that come out in the fission and therefore it was clear at the time that one of the ways to make possible the production of a chain reaction was to isolate the isotope U^{235} from the much more abundant isotope U^{238}.

Now, at present we have in our laboratories a row of bottles labeled, more or less, isotope—what shall I say—iron 56, for example, or uranium 235 or uranium 238 and these bottles are not quite as common as would be a row of bottles of chemical elements, but they are perfectly easily obtainable by putting due pressure on the Oak Ridge Laboratory (*laughter*). But at that time isotopes were considered almost magically inseparable. There was to be sure one exception, namely deuterium, which was already at that time available in bottles. But of course deuterium is an isotope in which the two isotopes hydrogen one and hydrogen two have a ratio of mass one to two, which is a very great ratio. But in the case of uranium the ratio of mass is merely 235 to 238, so the difference is barely over one percent. And that, of course, makes the differences of these two objects so tiny that it was not very clear that the job of separating large amounts of uranium 235 was one that could be taken seriously.

Well, therefore, in those early years near the end of 1939 two lines of attack to the problem of atomic energy started to emerge. One was as follows. The first step should be to separate in large amounts, amounts of kilograms or maybe amounts of tens of kilograms or maybe of hundreds of kilograms, nobody really knew how much would be needed, but something perhaps in that order of magnitude, separate such at that time fantastically large-looking amounts of uranium 235 and then operate with them without the ballast of the associated much larger amounts of uranium 238. The other school of thought was predicated on the hope that perhaps the neutrons would be a little bit more and that perhaps using some little amount of ingenuity one might use them efficiently and one might perhaps be able to achieve a chain reaction without having to separate the isotopes, a task as I say that at that time looked almost beyond human possibilities.

Now I personally had worked many years with neutrons, and especially slow neutrons, so I associated myself with the second team that wanted to use nonseparated uranium and try to do the best with it. Early attempts and studies, discussions, on how to separate the isotopes of uranium were started by Dunning and Booth in close consultation with Professor Urey. On the other hand, Szilard, Zinn, Anderson, and myself started experimentation on the other line whose first step involved lots of measurements.

Now, I have never yet quite understood why our measurements in those days were so poor. I'm noticing now that the measurements that we are doing on pion physics are very poor, presumably just because we have not learned the tricks. And, of course, the facilities that we had at that time were not as powerful as they are now. It's much easier to carry out experimentation with neutrons using a pile as a source of neutrons than it was in those days using radium-beryllium sources when geometry was the essential item to control or using the cyclotron when intensity was the desired feature rather than good geometry.

Well, we soon reached the conclusion that in order to have any chance of success with natural uranium we had to use slow neutrons. So there had to be a moderator. And this moderator could have been water or other substances. Water was soon discarded; it's very effective in slowing down neutrons, but still absorbs a little bit too many of them and we could not afford that. Then it was thought that graphite might be perhaps the better bet. It's not as efficient as water in slowing down neutrons; on the other hand little enough was known of its absorption properties that the hope that the absorption might be very low was quite tenable.

This brings us to the fall of 1939 when Einstein wrote his now famous letter to President Roosevelt advising him of what was the situation in physics—what was brewing and that he thought that the government had the duty to take an interest and to help along this development. And in fact help came along to the tune of $6000 a few months after and the $6000 were used in order to buy huge amounts—or what seemed at that time when the eye of physicists had not yet been distorted—(*laughter*) what seemed at that time a huge amount of graphite.

So physicists on the seventh floor of Pupin Laboratories started looking like coal miners (*laughter*) and the wives to whom these physicists came back tired at night were wondering what was happening. We know that there is smoke in the air, but after all . . . (*laughter*).

Well, what was happening was that in those days we were trying to learn something about the absorption properties of graphite, because perhaps graphite was no good. So, we built columns of graphite, maybe four feet on the side or something like that, maybe ten feet high. It was the first time when apparatus in physics, and these graphite columns were apparatus, was so big that you could climb on top of it—and you had to climb on top of it. Well, cyclotrons were the same way too, but anyway that was the first time when I started climbing on top of my equipment because it was just too tall—I'm not a tall man (*laughter*).

And the sources of neutrons were inserted at the bottom and we were studying how these neutrons were first slowed down and then diffused up the column and of course if there had been a strong absorption they would not have diffused very high. But because it turned out that the absorption was in fact small, they could diffuse quite readily up this column and by making a little bit of mathematical analysis of the situation it became possible to make the first guesses as to what was the absorption cross section of graphite, a key element in deciding the possibility or not of fabricating a chain reacting unit with graphite and natural uranium.

Well, I will not go into detail of this experimentation. That lasted really quite a number of years and required really quite many hours and many days and many weeks of extremely hard work. I may mention that very early our efforts were brought in connection with similar efforts that were taking place at Princeton University where a group with Wigner, Creutz and Bob Wilson set to work making some measurements that we had no possibility of carrying out at Columbia University.

Well, as time went on, we began to identify what had to be measured and how accurately these things that I shall call "eta," f, and p—I don't think I have time to define them for you—these three quantities "eta," f, and p had to be measured to establish what could be done and what could not be done. And, in fact, if I may say so, the product of "eta," f, and p had to be greater than one. It turns out, we now know, that if one does just about the best this product can be 1.1.

So, if we had been able to measure these three quantities to the accuracy of one percent we might have found that the product was for example 1.08 plus or minus 0.03 and if that had been the case we would have said let's go ahead, or if the product had turned out to be 0.95 plus or minus 0.03 perhaps we would have said just that this line of approach is not very promising, and we had better look for something else. However I've already commented on the extremely low quality of the measurements in neutron physics that could be done at the time—where the accuracy of measuring separately either "eta," or f, or p was perhaps with a plus or minus of 20 percent (*laughter*). If you compound, by the well-known rules of statistics, three errors of 20 percent you will find something around 35 percent. So if you should find, for example, 0.9 plus or minus 0.3—what do you know? Hardly anything at all (*laughter*). If you find 1.1 plus or minus 0.3—again, you don't know anything much. So that was the trouble and in fact if you look in our early work—what were the detailed values given by this or that experimenter to, for example, "eta" you find that it was off 20 percent and sometimes greater amounts. In fact I think it was strongly influenced by the temperament of the physicist. Shall we say optimistic physicists felt it unavoidable to push these quantities high and pessimistic physicists like myself tried to keep them somewhat on the low side (*laughter*).

Anyway, nobody really knew and we decided therefore that one had to do something else. One had to devise some kind of experiment that would give a complete over-all measurement directly of the product "eta," f, p without having to measure separately the three, because then perhaps the error would sort of drop down and permit us to reach conclusions.

Well, we went to Dean Pegram, who was then the man who could carry out magic around the University, and we explained to him that we needed a big room. And when we say big we meant a really big room, perhaps he made a crack about a church not being the most suited place for a physics laboratory in his talk, but I think a church would have been just precisely what we wanted (*laughter*). Well, he scouted around the campus and we went with him to dark corridors and under various heating pipes and so on to visit possible sites for this experiment and eventually a big room, not a church, but something that might have been compared in size with a church was discovered in Schermerhorn.

And there we started to construct this structure that at that time looked again in order of magnitude larger than anything that we had seen before. Actually if anybody would look at that structure now he would probably extract his magnifying glass (*laughter*) and go close to see it. But for the ideas of the time it looked really big. It was a structure of graphite bricks and spread through these graphite bricks in some sort of pattern were big cans, cubic cans, containing uranium oxide.

Now, graphite is a black substance, as you probably know. So is uranium oxide. And to handle many tons of both makes people very black. In fact it requires even strong people. And so, well we were reasonably strong, but I mean we were, after all, thinkers (*laughter*). So Dean Pegram again looked around and said that seems to be a job a little bit beyond your feeble strength, but there is a football squad at Columbia (*laughter*) that contains a dozen or so of very husky boys who take jobs by the hour just to carry them through College. Why don't you hire them?

And it was a marvelous idea; it was really a pleasure for once to direct the work of these husky boys, canning uranium—just shoving it in—handling packs of 50 or 100 pounds with the same ease as another person would have handled three or four pounds. In passing these cans fumes of all sorts of colors, mostly black, would go in the air (*laughter*).

Well, so grew what was called at the time the exponential pile. It was an exponential pile, because in the theory an exponential function enters—which is not surprising. And it was a structure that was designed to test in an integral way, without going down to fine details, whether the reactivity of the pile, the reproduction factor, would be greater or less than one. Well, it turned out to be 0.87. Now that is by 0.13 less than one and it was bad. However, at the moment we had a firm point to start from, and we had essentially to see whether we could squeeze the extra 0.13 or preferably a little bit more. Now there were many obvious things that could be done. First of all, I told you these big cans were canned in tin cans, so what has the iron to do? Iron can do only harm, can absorb neutrons, and we don't want that. So, out go the cans. Then, what about the purity of the materials? We took samples of uranium, and with our physicists' lack of skill in chemical analysis, we sort of tried to find out the impurities and certainly there were impurities. We would not know what they were, but they looked impressive, at least in bulk (*laughter*). So, now, what do these impurities do?—clearly they can do only harm. Maybe they make harm to the tune of 13 percent. Finally, the graphite was quite pure for the standards of that time, when graphite manufacturers were not concerned with avoiding those special impurities that absorb neutrons. But still there was some considerable gain to be made out there, and especially Szilard at that time took extremely decisive and strong steps to try to organize the early phases of production of pure materials. Now, he did a marvelous job which later on was taken over by a more powerful organization than was Szilard himself. Although to match Szilard it takes a few able-bodied customers (*laughter*).

Well, this brings us to Pearl Harbor. At that time, in fact I believe a few days before by accident, the interest in carrying through the uranium work was spreading; work somewhat similar to what was going on at Columbia had been initiated in a number of different Universities throughout the country. And the government started taking decisive action in order to organize the work, and, of course, Pearl Harbor gave the final and very decisive impetus to this organization. And it was decided in the high councils of the government that the work on the chain reaction produced by non-separated isotopes of uranium should go to Chicago.

That is the time when I left Columbia University, and after a few months of commuting between Chicago and New York eventually moved to Chicago to keep up the work there, and from then on, with a few notable exceptions, the work at Columbia was concentrated on the isotope-separation phase of the atomic energy project.

As I've indicated this work was initiated by Booth, Dunning, and Urey about 1940, 1939, and 1940, and with this reorganization a large laboratory was started at Columbia under the direction of Professor Urey. The work there was extremely successful and rapidly expanded into the build-up of a huge research laboratory which cooperated with the Union Carbide Company in establishing some of the separation plants at Oak Ridge. This was one of the three horses on which the directors of the atomic energy project had placed their bets, and as you know the three horses arrived almost simultaneously to the goal in the summer of 1945. I thank you. (*Applause*).

Chapter Four

Correspondence with Colleagues: Scientific, Political, Personal

In the archives of the University of Chicago there is a rich collection of Fermi's correspondence. In this chapter we have selected a few examples which cover a wide range of subjects and give one a broader sense of Fermi's character.

These letters date from the period when Fermi was professor of physics at the University of Chicago. One can notice that the speed of the mail service was much better in the 1950s than it is today. Fermi answered most promptly the letters that he felt important but often procrastinated in answering letters that were more in the category of a courtesy.

FERMI TO JAMES F. BYRNES, OCTOBER 16, 1945

This letter to Secretary of State James Byrnes indicates Fermi's deep interest in the recovery of Italy following the war. He is particularly concerned with the fate of the Jewish deportees from Italy.

October 16 1945

The Hon. James F. Byrnes
Secretary of State
Washington, D. C.

Dear Mr. Secretary:

There are two points that I would like to bring to
your attention concerning the relationships between the United
States and Italy. One is particular and one is general.

As you undoubtedly know as the Germans took over
complete control over Central and Northern Italy in September
1943 they arrested and deported to Germany a number of Italians
of jewish descent. Only a negligible fraction of these unfortunate
people has been able to come back to their homes after the end of
the war and the chances are that the great majority of them has
not survived. In spite of this it is natural that their families
should make all possible efforts to ascertain the fate of their
relatives.

I am informed that the Italian Committee for the
Jewish Deportees (Comitato Ricerche Deportati Ebrei) whose
chairman is Colonnel M. A. Vitale had succeeded in including
two of its officers in an Italian Military Mission bound for
Germany. The two officers had the assignment to find news of the
Italian deportees and to bring them assistance. It seems that
permission to cross the border has been denied to these officers
by the allied authorities.

I do not know of course what are the specific reasons
for this refusal. But it seems to me that if permission to go to
Germany could be granted to these two officers or to other representatives of the Committee one would not only contribute to a
humane endeavour but also contribute to the improvement of the
feelings of the Italian people towards the United States.

The second point is quite general. I do not expect
that what I am going to say is new to you; but my knowledge of
Italy where I was born and where I lived until seven years ago
may make it of some interest to you to hear my views.

I am convinced that the desire of the Italian people
to set up a democratic form of government is very real. Such
desire existed very strongly, though it was not expressed, at
the time when I was still living in Italy. I believe also that
the forces that tend to make the situation unstable are dangerously
strong. It seems to me to be in the interest of the United States
to encourage the democratic forces in Italy and to try to avoid
violent fluctuations of the public opinion that would make it
easy for extreme parties to destroy the present equilibrium.

2

From this point of view I believe that the peace treaty with Italy may be of great importance. A great number of Italians especially of the middle class who probably are one of the strongest democratic groups is exceedingly sensitive to questions that are deemed to reflect on the national honor. The final disposal of Trieste, to a lesser extent the disposal of the colonies and the admission of Italy to the United Nations will undoubtedly influence the state of mind of these Italians to an extent perhaps out of proportion with the importance of the actual issues involved. I believe that any help that the United States can give Italy in securing a satisfactory settlement of these problems would strengthen enormously the chances that Italy may settle in a stable democratic form of government.

Respectfully yours

Enrico Fermi

C. N. YANG TO FERMI, JANUARY 5, 1950

This exchange between Fermi and C. N. Yang concerns a numerical mistake in their joint paper "Are Mesons Elementary Particles?" (*Physical Review* 76 [1949]: 1739). Yang's letter goes on to discuss his work at the Institute for Advanced Study, principally concerned with properties of particles under space reflection. In a handwritten postscript Yang speaks of his fu-

THE INSTITUTE FOR ADVANCED STUDY
SCHOOL OF MATHEMATICS
PRINCETON, NEW JERSEY

January 5, 1950

Professor E. Fermi
Institute for Nuclear Studies
University of Chicago
Chicago 37, Illinois

Dear Professor Fermi:

I have made a check into the notes of our calculations about the structure of the meson and found that I had forgotten a factor $(2.03)^3 = 8.4$ in the process of computation so that the two constants in equation (18) are both too large by this factor, as you have already found out. Since this correction would make the value of $\overline{g^{12} / 4\pi\hbar c}$ about 70 times smaller than what we indicated in the paper I am extremely unhappy that I should have been so careless as to make such a mistake. Probably I should write an "Erratum" to the Phys. Rev. — I shall wait for your advice.

There is a typographical mistake in Equation (15) which should read

$$Q = 2i(f_1 + f_4)\,\sigma_1\,\sigma_2\,\sigma_3 \; -i(f_1 - f_4)\,\sigma_1\,\sigma_2\,\sigma_3\,\sigma_4 .$$

I don't believe I have reported to you about my work here since last November. Being aware of your kind interest I shall take this opportunity to do so.

(i) The problem concerning multiple meson production I eventually gave up for lack of any workable ideas. The problem is very interesting and I want to thank you for informing me about the discussions in Europe on the subject.

(ii) I noticed that while the reflection property of spin zero particles makes possible (or rather necessitates) the differentiation between scalar and pseudoscalar particles, similar differentiation of spin $\frac{1}{2}$ particles has not been discussed before. Upon closer examination it is found that four different inversion properties are possible for spin $\frac{1}{2}$ particles:

$$\psi \longrightarrow \beta\psi,\; -\beta\psi,\; i\beta\psi \quad \text{or} \quad -i\beta\psi.$$

An interesting consequence is that what we usually take as the scalar in the β-decay theory:

$$\psi_e^* \beta \psi_\nu$$

may actually be a pseudoscalar if e.g. under an inversion

$$\psi_e \longrightarrow \beta\psi_e ,\quad \psi_\nu \longrightarrow -\beta\psi_\nu .$$

ture and the possibility of returning to Chicago. Fermi writes at the bottom "offer highly probable." Fermi's response is to suggest additional ideas to be included in an erratum letter and assures Yang that a position for him at Chicago is in the works. Yang, however, remained at the Institute for Advanced Study. The paper "Are Mesons Elementary Particles?" was highly original and is referred to in Frank Wilczek's article in chapter 2.

THE INSTITUTE FOR ADVANCED STUDY
SCHOOL OF MATHEMATICS
PRINCETON, NEW JERSEY

[In fact in your original paper of 1934 on β-decay you did take $\psi_e^* \beta \psi_\nu$ as a pseudoscalar.]

Another interesting consequence arises in connection with the "universal" Fermi-type interaction. An additional selection rule which you suggested a year ago in a seminar [to prevent such disastrous reactions as

$$p^+ \longrightarrow \mu^+ + e^+ + \nu.]$$

can now be made a natural consequence of the proposal of invariance under inversion by assuming e.g.

$$\psi_\mu \rightarrow \beta \psi_\mu, \ \psi_e \rightarrow \beta \psi_e, \ \psi_\nu \rightarrow \beta \psi_\nu, \ \psi_p \rightarrow i\beta \psi_p, \ \psi_\eta \rightarrow i\beta \psi_\eta.$$

This assignment is, however, not completely satisfactory. I am writing a paper with Tiomno on all these things.

(iii) The problem of the possibility of integration in Heisenberg's representation I have pushed further. However, I have not completely succeeded in identifying it with Feynman's theory in all cases. My present state of progress is described in a letter to Kallen, a copy of which I am enclosing.

(iv) Under some pressure from Marshak who says that they will soon try the experiment of differentiating between scalar and pseudoscalar neutral mesons, I have made some more detailed calculations about the expected difference. This was done last week and if I did not make a mistake the experiment will not be as difficult as a rough estimate would indicate it to be. I shall send you my results as soon as they are ready.

Please give my best regards to Mrs. Fermi, Nella and Julio (do I spell his name right?).

Yours very sincerely,

Frank

C. N. Yang

CNY:jj

P.S. I recently got some letters from my father. He strongly discourages me to go back to China. So I think I would want to be back at Chicago next fall. Do you think I should write immediately to Dr. Allison?

Offer highly probably   Frank

FERMI TO C. N. YANG, JANUARY 12, 1950

January 12, 1950

Dr. C. N. Yang
Institute for Advanced Studies
Princeton, New Jersey

Dear Yang:

I have delayed answering your letter of January 5 because I was proposing to look into a number of points, namely:

1. What is the situation if one assumes a vector meson instead of a pseudo-scalar meson?

2. In the case of the pseudo-scalar meson what happens if one takes the basic interaction as a tensor?

and a few possibilities of this same type.

It seems to me that before writing a letter to correct our previous mistake we should have explored what these possibilities are. I have not found so far the time to do so and I wonder whether you would be in a position to look into some of these points.

Another point which I believe should be explored is the following:

The pseudo-scalar interaction of the pseudo-scalar meson does not give in second approximation static nuclear forces. I believe that it is supposed to do so in fourth approximation but I am not clear as to what size of the coupling coefficient is needed in order to obtain force of a plausible intensity. This last is perhaps a point that you can clear merely by talking to some of the specialists around the Institute.

I was glad to hear that you now plan to come back to Chicago next year although I realize that this may be for you a disappointment. Already some time ago we have set in motion a machinery to get a proper appointment for you and although no final word has yet been obtained from the administration, Allison tells me that he is very optimistic as to the outcome. There is no need for you to write to Allison because I have already informed him of what you wrote to me.

Sincerely yours,

EF:HL   Enrico Fermi

ERWIN SCHRÖDINGER TO FERMI, FEBRUARY 10, 1951

This exchange between Erwin Schrödinger and Fermi concerns a loan of 400 lire that Fermi made to Schrödinger in 1938. Schrödinger, penniless, left Austria for Rome in September 1938 to escape arrest by the Nazis. As he passed through Rome, Fermi made the loan that is the subject of the exchange of letters. (A description of Schrödinger's passage through Rome can be found in a recent biography, Walter Moore, *Schrödinger: Life and Thought* [Cambridge University Press, 1989], p. 344.) Schrödinger expresses a feel-

10. II. 1951.

26 KINCORA ROAD
CLONTARF,
DUBLIN

Dear Fermi,

Returning to Dublin from nearly 6 months absence on the Continent (mainly at Innsbruck) I find your and Sra Fermi's kind new years greetings to myself and Mrs. Schrödinger (I really wanted to write: and my wife, to get away from the formal Anglo-Saxon style), which we heartily return.

We are thinking of you very often, and with regret that we have not met for so many years. This can perhaps not be helped at the moment, though I hope it will be helped once. For the moment however I beg you to help me remove once and for all a remorse that I cannot help associating with my memory of you and our last meeting, namely that I still owe you Lire 400 val. Sept. 1938. To re-calculate this sum to date, now that all money-value has gone down, is very difficult, but I think that something like 200 Swedish Crowns would be a modest estimate for re-payment. If you agree and if you still have an account at Stockholm, this would be very simple. If the latter is not the case, please indicate me your bankers' account at Chicago, and I hope to manage even so.

I enjoyed my long stay in Tirol very much. I

ing of "remorse" that he has not been able to settle the debt. But the letter also gives a somber view of the intellectual (spiritual) side of postwar Austria. Fermi's response is warm and cordial, but one can note that he writes, "if world conditions will permit." This expression of concern for world peace shows up in several of Fermi's letters and speeches. Schrödinger writes of being in the "prime of . . . elderlyness" at the age of sixty-three.

The legibility of Schrödinger's writing is difficult, and the text has been retyped.

had almost forgotten what a real mountain looks like, and I was happy to be able to stay long enough for recovering practice in moderate walking tours – at least in summer. My efforts to take up ski-ing were not so very successful, the time was too short for that, and the new technique of 'parallel-swings', to which the Kandahar-binding (Binding) is singularly adapted, is a thing for vigorous youths but not for persons "in the prime of their elderlyness" like myself.

With regard to all material needs Austria is at present a cheap and rich country. It was over-crowded by foreigners both during the summer and the winter season. There is no rationing of anything and the shops overflow with lovely goods. Against this the spiritual austerity is appalling. Foreign books and periodicals are practically beyond reach, except perhaps in a few big Viennese Institutes. The allowance of a provincial Institute of Theoretical Physics is of the order of one or two thousand Austrian shillings p.a. (meaning 40 or 80 dollars), simply ridiculous. The most common and urgently needed text-books are unobtainable.

With best wishes and kindest regards from my wife and myself to both of you.

Yours very sincerely

E. Schrödinger

February 10, 1951

Dear Fermi,

Returning to Dublin from nearly 6 month's absence on the Continent (mainly at Innsbruck) I find your and S^{ra} Fermi's kind new years greetings to myself and Mrs. Schrödinger (I really wanted to write: and my wife, to get away from the formal Anglosaxon style), which we heartily return.

We are thinking of you very often, and regret that we have not met for so many years. This can perhaps not be helped at the moment, though I hope it will be helped once. For the moment however I beg you to help me remove once and for all, a remorse that I cannot help associating with my memory of you and our last meeting, namely that I still owe you Lire 400 val. Sept 1938. To re-calculate this sum to date, now that all money-value has gone down, is very difficult, but I think that something like 200 Swedish Crowns would be a modest estimate for re-payment. If you agree and if you still have an account at Stockholm, this would be very simple. If the latter is not the case, please indicate me your bankers' account at Chicago, and I hope to manage even so.

I enjoyed my long stay in Tirol very much. I had almost forgotten what a real mountain looks like, and I was happy to be able to stay long enough for recovering practice in moderate walking tours—at least in the summer. My efforts to take up ski-ing were not so very successful, the time was too short for that, and the new technique of "parallel-swings," to which the Kandahar-binding (Bindung) is singularly adapted, is a thing for vigorous youths but not for persons "in the prime of their elderlyness" like myself.

With regard to all the material needs Austria is a cheap and rich country. It was over-crowded by foreigners both during the summer and the winter season. There is no rationing of anything and the shops overflow with lovely goods. Against this, the spiritual austerity is appalling. Foreign books and periodicals are practically beyond reach, except perhaps in a few big Viennese Institutes. The allowance of a provincial Institute of Theoretical Physics is of the order of one or two thousand Austrian schillings p. a. (meaning 40 or 80 dollars), simply ridiculous. The most common and urgently needed textbooks are unobtainable.

With best wishes and kindest regards from my wife and myself to both of you.

Yours very sincerely,

E. Schrödinger

FERMI TO ERWIN SCHRÖDINGER, FEBRUARY 27, 1951

February 27, 1951

Dr. E. Shrodinger
26, Kingora Road
Clontarf
Dublin, Ireland

Dear Shrodinger:

It was very nice to hear from you and I am glad to know that
Mrs. Shrodinger and you have been able to spend six months in your
home country.

As to the old debt that you mention, I believe that you are estimating
the value of the 400 lire too high. At that time the lire was worth
about one twentieth of one dollar and it seems that therefore a $20.00
settlement would be correct. I no longer have an account in Sweden. My
bank here in Chicago is the University National Bank, 1354 East 55th
Street, Chicago 15. Please, however, be sure if there are any difficulties whatsoever about transferring this amount not to worry about
it because it is certainly not worth it.

I have been in Italy once since the war and I had the pleasure to find
that at least physics is as prosperous as the circumstances warrant.

We have just moved into a new Institute building here in Chicago and we
have a 450 Mev cyclotron which has started its initial operation just
in these days. If world conditions will permit, we hope to be able to
do something with it.

With best wishes to you and to Mrs. Shrodinger also from my wife.

                                        Yours very sincerely,

EF:HL   Enrico Fermi

FRED REINES AND CLYDE COWAN TO FERMI, OCTOBER 4, 1952

This exchange between Fred Reines and Clyde Cowan and Fermi concerns their plans for an experiment to directly detect neutrinos. The letter announces that they will be able to use reactor neutrinos instead of neutrinos from a nuclear explosion. (The designation "5819" refers to a particular RCA photomultiplier.) The Reines-Cowan experiment was successfully performed in 1953 and published in *Physical Review* 92 (1953): 830.

Los Alamos
October 4, 1952

Dear Enrico,

We thought that you might be interested in the latest version of our experiment to detect the free neutrino, hence this letter. As you recall, we planned to use a nuclear explosion for the source because of background difficulties. Only last week it occurred to us that background problems could be reduced to the point where a Hanford pile would suffice by counting only delayed coincidences between the positron pulse and neutron capture pulse. You will remember that the reaction we plan to use is $p + \nu \rightarrow n + \beta^{+}$. Boron loading a liquid scintillator makes it possible to adjust the mean time τ between these two events and we are considering $\tau \sim 10\,\mu$sec. Our detector is a ten cubic foot "square" fluor filled cylinder surrounded by about 90 5819's operating as two large tubes of 45 5819's each. These two banks of ganged tubes isotropically distributed about the curved cylindrical wall are in coincidence to cut tube noise. The inner wall of the chamber will be coated with a diffuse reflector and in all we expect the system to be energy sensitive, and not particularly sensitive to the position of the event in the fluor. This energy sensitivity will be used to discriminate further against background. Cosmic ray anticoincidence will be used in addition to mercury or adanac lead for shielding against natural radioactivity. We plan to immerse the entire detector in a large borax water solution for further necessary reduction of pile background below that provided by the Hanford shield.

- 2 -

Fortunately, the fast reactor here at Los Alamos provides the same leakage flux as Hanford so that we can check our gear before going to Hanford. Further, if we allow enough fast neutrons from the fast reactor to leak into our detector we can simulate double pulses because of the proton recoil pulse followed by the neutron capture which occurs in this case. We expect a counting rate at Hanford in our detector about six feet from the pile face of ~1/5/min. with a background somewhat lower than this.

As you can imagine we are quite excited about the whole business, have cancelled preparations for use of a bomb, and are working like mad to carry out the ideas sketched above. Because of the enormous simplification in the experiment we have already made rapid progress with the electronic gear and associated equipment and expect that in the next few months we shall be at Hanford reaching for the slippery particle.

We would of course appreciate any comments you might care to make.

Sincerely yours,

Fred Reines
Clyde L. Cowan

FR:daf

cc: File

FERMI TO FRED REINES, OCTOBER 8, 1952

October 8, 1952

Dr. Fred Reines
Los Alamos Scientific Laboratory
P.O. Box 1663
Los Alamos, New Mexico

Dear Fred:

Thank you for your letter of October 4th by Clyde Cowan and
yourself. I was very much interested in your new plan for
the detection of the neutrino. Certainly your new method
should be much simpler to carry out and have the great advantage that the measurement can be repeated any number of
times. I shall be very interested in seeing how your 10 cubic
foot scintillation counter is going to work, but I do not know
of any reason why it should not.

Good luck.

                                    Sincerely yours,

                                    Enrico Fermi

EF:vr

FERMI (AND OTHER MEMBERS OF THE INSTITUTE FOR NUCLEAR STUDIES) TO DEAN G. ACHESON, MAY 22, 1952

A letter to Secretary of State Dean Acheson from six prominent members of the Institute for Nuclear Studies, including Fermi, protesting the refusal of the government to grant a passport to Linus Pauling for travel to a meet-

THE UNIVERSITY OF CHICAGO

CHICAGO 37 · ILLINOIS

INSTITUTE FOR NUCLEAR STUDIES

May 22, 1952

The Honorable Dean G. Acheson
Secretary of State
Washington 25, D. C.

My dear Mr. Secretary:

The scientific community has been deeply shocked by the statement that Professor Linus Pauling, American citizen, former President of the American Chemical Society, and Professor of Chemistry at the California Institute of Technology has been refused an American passport for travel to a meeting of the Royal Society in England.

Most of the scientists of America feel that they know Linus Pauling. All of us know his scientific work. Many of us who sign this letter have known him with varying degrees of intimacy for very many years, and have discussed with him not only scientific matters but subjects of social and political import. Probably none of us have agreed with him in all subjects. We have all found reason to respect his scientific opinions. Most, if not all of us, have disagreed with his political views, and some of us may have found occasion to regard them as not even worthy of the high regard we have found necessary for his scientific viewpoint. But no one of us, nor any with whom we have talked, has questioned his integrity and sincerity nor his essential loyalty to the United States.

We have all been long convinced that the world is deeply enmeshed in a desperate struggle between the forces of evil repression and those of freedom and liberty. Those of us who sign this letter have had no doubt that the Russian Government has represented the evil in this conflict. We have always hoped that the Government of the United States would stand clearly on the side of freedom and liberty. We cannot reconcile with this hope the withdrawal of passport privileges without trial, without a hearing, and without recourse to an appeal, on the suspicion that the political opinions of a citizen are not those of the majority of the nation.

ing of the Royal Society in England. As a result of this letter and others Linus Pauling was issued a short-term passport to attend the Royal Society meeting in the summer of 1952. However, he was denied a permanent passport over the next several years.

Page 2
May 22, 1952

Our main plea in this case is a matter of pure principle. The principle of our freedom is that a man is innocent of wrongdoing until proven guilty before a jury of his peers of violation of a law of our land. Professor Pauling is a man of international fame. It is essential for the maintenance of what remains of our free world that free interchange of ideas within this world be sustained. The travel of citizens of the free countries within this free community is an absolutely essential requirement of this freedom. Since Professor Pauling is not guilty of violation of our laws we see no justification for suspension of this fundamental necessary freedom.

But even aside from our interest in the principle of freedom we are incredulous of the reason given for withholding the passport: that it "is not in the best interests of the United States" to grant the passport. We cannot believe, with the greatest stretch of our imagination, that any reason can exist which would make the granting of a passport of so great harm to this country as its withdrawal.

We are well aware that information, or the suspicion of information, of which we are not informed may be in governmental files. Those of us who think that we know Pauling well cannot believe that any really detrimental facts exist. Those of us who know him less well may be less sure of this. But none of us can imagine circumstances by which the granting of a passport can be one-tenth so harmful to the interests of this country as the creation in the world of this "cause celebre" that its withholding has done. What harm, what information, what tales could Professor Pauling take with him to England, even were he so inclined, that can compare in damage to the incredible advertisement that this country forbids one of its most illustrious citizens to travel? We ask you respectfully, Mr. Secretary, that this decision be reversed.

Very truly yours,

Samuel K. Allison   Willard F. Libby

Herbert L. Anderson   Joseph E. Mayer

Enrico Fermi   Harold C. Urey

LINUS PAULING TO FERMI, JUNE 16, 1952

Pauling's response to Fermi thanking him for his support.

CALIFORNIA INSTITUTE OF TECHNOLOGY
PASADENA 4

GATES AND CRELLIN LABORATORIES OF CHEMISTRY

16 June 1952

Professor Enrico Fermi
Institute for Nuclear Studies
University of Chicago
Chicago 37, Illinois

Dear Enrico:

I am grateful to you for having signed the letter of
22 May to the Secretary of State, a copy of which has just
reached me.

I think that it is an extraordinarily fine letter, and
that the Secretary of State can hardly ignore the arguments
that you have presented.

I agree with practically everything in the letter. Especially, I have felt sure that the refusal of a passport to
me would reflect discredit upon our Government, and it was in
part for this reason that I tried very hard to get the original
decision reversed.

With best regards, I am

Sincerely yours,

Linus Pauling:W

GEORGE GAMOW TO FERMI, MARCH 13, 1953

This exchange of letters between George Gamow and Fermi concerns a popular article by Gamow. Fermi takes pains to correct Gamow's Italian. Reference to this article is made at the conclusion of M. L. Goldberger's article in chapter 6.

GEORGE GAMOW
19 THOREAU DRIVE
BETHESDA, MARYLAND

March 13th
1953

Dear Enrico,

I have just written an article on the elementarity of elementary particles to be published probably in Scientific American. Since it involves you, and your π-production theory, I would like very much to have it checked by you.* Just make your remarks, and corrections on the included copy, and send it back. In particular: is my italian spelling O.K. on p. 14.? I do not have the copies of the diagrams, but you can easyly immagine them.

We are making some progress with turbulent distribution of galaxies, but it is a lot of numerical work!

Looking forward to see you in Chicago on Apr. 25 (if you will be there) Yours as ever   Geo.

P.S. Did you like my "New Genesis"?

*Please make allowence for layman's presentation.

FERMI TO GEORGE GAMOW, MARCH 24, 1953

March 24, 1953

Dr. George Gamow
19 Thoreau Drive
Bethesda, Maryland

Dear George:

The incident that you relate on page 14 of your manuscript is
essentially true except for a few mistakes in the personalities
and in Italian. As far as I recall, the exalted person who
attended the academy meeting was not Mussolini but the King.
Furthermore, the Italian for "car driver" is not "cocchiere"
which means driver of a carriage with horses, but "chauffeur".
Also "eccellenza" is feminine, therefore, it should be "Sono
lo chauffeur di sua Eccellenza Fermi".

I enjoyed your "New Genesis".

                                        Sincerely yours,

                                        Enrico Fermi

EF:vr

SAMUEL GOUDSMIT TO FERMI, MARCH 11, 1953

An exchange between Sam Goudsmit and Fermi, concerning the term "isotopic spin." The term was a compromise. Today it has become "isospin," a possibility not considered in this exchange.

THE PHYSICAL REVIEW

PUBLISHED BY THE AMERICAN INSTITUTE OF PHYSICS FOR THE

AMERICAN PHYSICAL SOCIETY

S. A. GOUDSMIT, EDITOR
S. PASTERNACK, ASSISTANT EDITOR

BROOKHAVEN NATIONAL LABORATORY
UPTON, LONG ISLAND, N.Y.
PATCHOGUE 2600, EXT. 2341

11 March 1953

Dr. Enrico Fermi
Institute for Nuclear Studies
University of Chicago
Chicago 37, Illinois

Dear Enrico:

I heard, from Telegdi, and notice also in the paper by you and your colleagues that you are in favor of abbreviating "isotopic spin" to "ispin" and that this abbreviation has the support of the whole Institute for Nuclear Studies.

I would like to voice my disapproval of this abbreviation. The Physical Review work has taught me that almost every author tries to invent new words and new abbreviations, which are not always wisely chosen. I quite agree that the name "isotopic spin" was originally a very bad choice for that concept, but I don't think that abbreviating it to "ispin" makes it any better. It might even cause confusion since the letter "i" or "I" is generally used for the nuclear spin. I don't even know how to pronounce it. Would you make it "ice-pin" or do you call it "is-pin". I do not think this abbreviation, nor the many others we receive for all sorts of physical quantities, serves any useful purpose.

However, if you want to get away from the meaningless expression "isotopic spin" I would strongly suggest that, from now on, it be denoted by the non-committal expression "T-spin" or "γ-spin". We are changing the notations in your and in Telegdi's paper to "T-spin" so please let us know at once in case you do not agree with this change, so that publication will not be unnecessarily delayed.

Do you know anything more about your plans for next June.

Best regards.

Sincerely yours,

Sam

S. A. Goudsmit

cc: V. L. Telegdi

FERMI TO SAMUEL GOUDSMIT, MARCH 24, 1953

March 24, 1953

Dr. Samuel Goudsmit
Brookhaven National Laboratory
Upton, Long Island, New York

Dear Sam:

I am sorry to hear that you dislike the expression "ispin". I must say, on the other hand, that I dislike "T-spin" and I would suggest as a compromise that we go back temporarily to the traditional misnomer "isotopic spin" and we leave to the future to settle the difference.

Concerning my visit to Brookhaven, it appears now that no member of my family, except perhaps Laura will accompany me. I mentioned to her that you probably had to know her decision before too long and she said that if she had to decide now, she would not come. Therefore, if you can wait a little longer for her decision, there is a chance that she may change her mind, but if you must have the answer right away in order to arrange for housing, count on me only.

Sincerely yours,

Enrico Fermi

EF:vr

P.S. Telegdi would also like to have the expression "isotopic spin" substituted in his paper with Gell-Mann for "ispin".

GEORGE KISTIAKOWSKY TO FERMI, SEPTEMBER 29, 1953

An exchange between Fermi and George Kistiakowsky requesting a recommendation of Maria Mayer for a distinguished appointment (presumably at Harvard). Fermi's letter is a brief but powerful recommendation. The rapid response of Fermi demonstrates his enthusiasm for Maria Mayer and, incidentally, the efficiency of the post in those days.

HARVARD UNIVERSITY
DEPARTMENT OF CHEMISTRY

12 Oxford Street
Cambridge 38, Massachusetts, U.S.A.
September 29, 1953

Professor Enrico Fermi
Institute for Nuclear Studies
University of Chicago
Chicago 37, Illinois

Dear Enrico:

I had the misfortune of having been made a member of a committee which is recommending a woman scholar for a distinguished appointment. There is no limitation on the field and since the physical scientists are heavily outnumbered by representatives of other scholarly fields, the probability is that the final choice will not be a physical scientist. Nonetheless, I have put forward the candidacy of Maria Mayer, and I am wondering whether it would be an imposition to ask you for a confidential letter with an expression of your opinion of her as a scientist and as a teacher. As the next and final meeting of the committee is on November 3, I would need the letter a few days ahead of that date.

With best personal regards,

Sincerely yours,

George

G. B. Kistiakowsky

GBK:EW

FERMI TO GEORGE KISTIAKOWSKY, SEPTEMBER 30, 1953

September 30, 1953

Professor G. B. Kistiakowsky
Department of Chemistry
Harvard University
Cambridge 38, Massachusetts

Dear George:

I believe that Maria G. Mayer is without any qualification an
outstanding scientist. She began her career in physics with
work on the quantum theory of radiation, but soon her interest
shifted to the applications of quantum mechanics to molecular
structure. Her work on the theory of the benzene ring is now
classical.

More recently she has made fundamental advances in the understanding of the structure of the nucleus with her shell model
with strong spin orbit coupling. This simple idea has been
basic in the interpretation of nuclear phenomena and is
doubtlessly the most important contribution to the physics of
the nucleus in the last five years.

Sincerely yours,

Enrico Fermi

EF:vr

ARTHUR COMPTON TO FERMI, DECEMBER 2, 1953

An exchange of letters between Arthur Compton and Fermi on the occasion of the eleventh anniversary of the first chain reaction. There is a sharp contrast between Compton's optimism and Fermi's more guarded and pessimistic view. In Fermi's reply he refers to the "Atoms for Peace" speech given by President Eisenhower on December 8, 1953, to the United Nations.

WASHINGTON ⚘ UNIVERSITY
SAINT LOUIS
OFFICE OF THE CHANCELLOR

Dec. 2, 1953

Prof. Enrico Fermi,
Chicago

Dear Enrico:

I can't let December second
pass by without recalling the great
experiment of eleven years ago,
in which you played the key rôle.
What its significance will eventually
be, how can one tell? Yet it's
my own definite belief that our
work on that occasion has had
a favorable effect on stabilizing
the world in favor of freedom.

Ferm. vol. 3

When we shall reap rewards in terms of power available when it would not otherwise be, I'm not now guessing.

But enough of this. I just wanted to pass on to you, and to our other friends at Chicago, our hot greetings – especially to Laura

Yours sincerely,
Arthur Compton

FERMI TO ARTHUR COMPTON, DECEMBER 14, 1953

December 14, 1953

Professor Arthur Compton
Office of the Chancellor
Washington University
St. Louis, Missouri

Dear Arthur:

Coming back to Chicago after two weeks at Harvard, I find
your kind note of December 2nd. Perhaps the hope that you
express has become a little bit more concrete after the
recent speech of the President, however, it is difficult
to predict the future.

Please accept my best wishes for Mrs. Compton and yourself.

Sincerely yours,

Enrico Fermi

EF:vr

OWEN CHAMBERLAIN TO FERMI FEBRUARY 2, 1954

A letter from Owen Chamberlain to Fermi. The first paragraph refers to information about measurements of proton polarization at the Berkeley cyclotron by Chamberlain and Emilio Segrè. The reason for including this letter is the final paragraph with Chamberlain's moving reference to Fermi as a mentor.

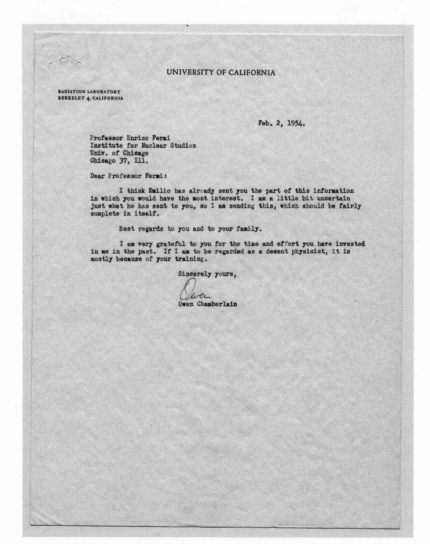

UNIVERSITY OF CALIFORNIA

RADIATION LABORATORY
BERKELEY 4, CALIFORNIA

Feb. 2, 1954.

Professor Enrico Fermi
Institute for Nuclear Studies
Univ. of Chicago
Chicago 37, Ill.

Dear Professor Fermi:

I think Emilio has already sent you the part of this information in which you would have the most interest. I am a little bit uncertain just what he has sent to you, so I am sending this, which should be fairly complete in itself.

Best regards to you and to your family.

I am very grateful to you for the time and effort you have invested in me in the past. If I am to be regarded as a decent physicist, it is mostly because of your training.

Sincerely yours,

Owen Chamberlain

Chapter Five

Research and Teaching: Selections from the Archives

INTERNATIONAL HOUSE APPLICATION, JUNE 10, 1940

Fermi made a number of trips to Chicago. On several of these trips he stayed at the International House. Fermi dutifully fills out an application that is hardly designed for a man of his stature. In June 1940 Fermi was the guest of Arthur Compton. On this occasion Fermi gave lectures on cosmic rays. This visit had nothing to do with uranium research (Arthur H. Compton, *Atomic Quest* [Oxford University Press, 1956], p. 43.) Subsequent stays at International House were in the winter and spring of 1942, when Fermi was following the move of the pile research to the University of Chicago.

I.H. Form No.3

INTERNATIONAL HOUSE
1414 E 59th St., Chicago

APPLICATION FOR ADMISSION

Resident Membership   (X)
Non-Resident Membership   ( )

Approved

SPECIFY QUARTER
☐ Autumn   Quarter 19
☐ Winter   Quarter 19
☐ Spring   Quarter 19
☒ Summer   Quarter 19
   ☒ 1st term
   ☐ 2nd term

1. Name (Print) Mr. Miss Mrs. ENRICO   FERMI
   (First)   (Middle)   (Last)

2. Present mailing address (Print) 382 Summit Ave. Leonia N.J. Telephone Leonia 4-2710
   (Street and Number)   (City)   (State)
   Home address (Print) 382 Summit Ave. Leonia   N.J.
   (Street and Number)   (City)   (State)
   Name and address of parents (or nearest relative) Laura Fermi (Wife)
   382 Summit Ave. Leonia N.J.

3. Place of birth Rome, Italy   Date of birth September 29, 1901
   Country of birth of Mother Italy   Country of birth of Father Italy

4. Now a citizen of what country? Italy   Nationality Italy
   If foreign-born, when did you arrive in the United States? January 2 1939
   Probable date of departure Permanent resident Have you become, or do you intend to become
   a U. S. citizen? yes   If so, when? as soon as possible

5. If American, list countries visited and dates

6. With what educational institution in Chicago will you will be affiliated during the period for which you seek
   membership in International House? University of Chicago
   Specify status Faculty   Check:
   (as 1st 2nd 3rd, or 4th year undergraduate, graduate, faculty or staff)   Full time (V) Part-time ( )
   Are you enrolled in that school at present?   If not, when will you be so enrolled?
   Will you seek a degree? no If so, which degree?   Department of specialization Physics

7. List in chronological order all schools attended, including professional schools, and present college:

Name and location of institution	Dates of attendance	Degree or Diploma and date	Name of Dean or Department Head
University of Pisa Italy	1918-22	Ph.D 1922	L. Puccianti

8. Please give two references.
   Name   Address   Position
   1. Arthur H. Compton   University of Chicago   Professor
   2.

9. Have you been a member of either the New York or Berkeley International Houses? I have elected
   Which House? Berkeley   If so, when? February 1940

—1—

10. What business and professional experience have you had? List below.

Firm, school, or organization, and location	Kind of work	Inclusive dates
University of Rome, Italy	*Professor*	*1927–38*
Columbia University, New York	*Professor*	*1939 –*

11. Married (*X*) Single ( )

12. Give concise information concerning articles or books published (title, when, where, and by whom published), research, inventions, or any other creative work. *Several papers on physics in specialized journals*

13. What special recognition, if any, have you received for excellence in academic work, such as honors, prizes, scholarships? *Nobel prize for Physics 1938*

14. What languages do you speak fluently? *English, Italian, German, French* Read? *same*

15. Religious affiliation or preference *Catholic* If none, give religious affiliation or preference of your parents: Mother *Catholic* Father *Catholic*

16. Credit Reference:

    (a) Who is responsible for your indebtedness? *(University Branch 113th Street & Broadway, New York*

    (b) If you are fully liable, give name and location of your Bank *Corn Exchange Bank)*

    (c) If support is dependent upon a loan, have you been assured that one is available?

17. Check thus (X) the amount of time you expect to give to outside employment during the coming year: None (*X*) Part-time ( ) Full time ( ) Have you a definite promise of employment? *yes* With whom? *Columbia University, New York* Indicate nature of employment obtained or desired *Professor*

18. What influenced you to apply for membership in International House?

Letter from Mr. E. B. Price of May 14 1940

—2—

International House is an educational enterprise which is devoted to the promotion of international friendships and understanding through social, intellectual, spiritual, and physical activities on the part of the members of the House. Membership is limited to students, faculty, or staff members in the higher educational institutions in Chicago and vicinity, regardless of religion, nationality, race, color, or sex.

The undersigned is interested in the objects of International House above set forth and agrees as a condition of membership in International House, whether as a resident or non-resident member, to participate in such activities and to join with the officers and members of the House in developing them. It is understood that this membership is issued in conformity with the rules and regulations of International House and that all privileges and benefits conferred by it are subject to revocation by the officers of International House.

I certify that the information on pages 1 to 3 is complete and accurate.

Signature _Enrico Fermi_

Date _June 10 1940_ 193___

To those applying for resident membership: Please read this section carefully and fill it in completely.

A deposit of $10.00 must accompany applications for Resident Membership.

State maximum price and location of room desired _7.50 per week_

Rooms range in price from $5.00 per week on the second floor to $7.50 per week on the top floors. Some women's rooms have running water. If you desire this accommodation, please state. There are a few double rooms; and also several rooms with attached bath, at $10.00 a week.

First Choice _Preferably on a top floor_

Second Choice _____

Deposits are refundable when the House is not able to assign a room within the price range specified by an applicant, or when an applicant notifies the House at least fifteen days prior to the opening of the Quarter, that he wishes to cancel the application.

The applicant should not write below this line:

Activities Card executed_____Not executed_____

—3—

FERMI TO DEAN WALTER BARTKY, DECEMBER 3, 1945

In July 1945 Walter Bartky, dean of the Division of Physical Sciences at the University of Chicago, visited Fermi at Los Alamos. The purpose was to invite Fermi to return to Chicago as professor of physics and become a member of a newly formed Institute for Nuclear Studies. Fermi accepted and attracted to the institute an extraordinary group of scientists. In the following letter, written at the end of 1945, Fermi outlines to Bartky the program of

December 3, 1945

TO: Walter Bartky

FROM: E. Fermi

SUBJECT: 100 Mev Betatron

   In formulating a program for the research work in experimental physics at the Institute of Nuclear Studies, we should consider two main directions of our activities; namely, (A) research in the field of excitation energies of the order of magnitude up to 20 Mev and (B) research in the field of very high excitations.

   There is no doubt that a very great amount of work remains to be done in the first of these fields. This covers, among other topics, the study of the physical properties of neutrons which will probably be investigated with the use of piles with an accuracy much greater than has hitherto been possible. Also the use of other machines that are available to the staff of the Institute, notably the cyclotron and the D-D source, will undoubtedly prove of great value for these investigations. I believe that it is fair to look towards many years of fruitful research that will be made possible by the use of these devices.

   I share, on the other hand, the rather wide spread opinion that the best chance for the investigation of the most fundamental properties of the atomic nuclei will be offered by the use of machines capable of reaching up to energies of the order of 100 Mev. It is not possible to make any too definite predictions as to what one might expect; indeed only very little information is available as to the properties of high energy particles, and this has mostly been obtained from cosmic ray work. There are, however, some theoretical points of view which offer a fairly reliable clue. It is generally believed that forces between nuclear particles are transmitted by the so-called mesotron field. This is a field similar in some of its features to the electromagnetic field which is responsible for the electric interaction between charged particles. In particular, quanta of the mesotron field; namely, mesotrons, can be emitted by a sufficiently highly excited nuclear system in the same way as quanta of the electromagnetic field; namely, photons or light quanta, can be emitted by an excited electronic system. A fundamental difference consists in the fact that light quanta have no intrinsic energy content so that they can be emitted even when the excitation energy is very low. This is not true of the mesotrons which are believed to have an intrinsic energy of the order of 80 Mev and cannot, therefore, be emitted unless at least this amount of excitation energy is available in the nuclear system.

   It should be realized, of course, that the theory of nuclear forces here outlined is highly speculative. Its experimental basis is to be found primarily in two facts which are far from being conclusive. One is that mesotrons having properties apparently fairly similar to those that are postulated in the theory of nuclear forces are observed as a common component of cosmic radiation. The second point is that the observed range of the nuclear forces is satisfactorily explained by the mesotron theory.

   The possibility of exciting atomic nuclei to such energies that mesotrons may be emitted constitutes one of the most fascinating possibili-

research that he expects for the institute. Fermi presents a very clear view that the most important area of research is the frontier opened by observations in cosmic radiation. He clearly is looking ahead to new areas beyond the Los Alamos experience. The research was aimed to explore the highest-energy excitations and the artificial production of mesotrons. At the time the mesotron was the mu-meson; pi-mesons were yet to be discovered.

Walter Bartky   -2-   December 3, 1945

ties of the high energy field. If mesotrons could be produced in the laboratory, it would seem likely that their study and the consequent understanding of nuclear forces would progress at an enormously faster rate than is made possible by the accumulation of evidence on cosmic ray particles. Even apart from these possibilities there are a great number of interesting objectives for the physics of high nuclear excitations. Since the average binding energy of protons and neutrons inside the nucleus is of the order of magnitude of 10 Mev, it is fair to expect that a neutron excited to an energy of the order of 100 Mev will literally go to pieces, yielding a large number of new radioactive species. The investigations of these transmutation phenomena of much greater complexity than any that have been known until now will offer an almost unlimited field of investigation, both for the physicist and for the chemist.

The devices that are capable of producing particles accelerated in the 100 Mev region fall into two groups dependent on whether heavy particles or electrons are accelerated. The main attempt at producing high energy heavy particles is now being made at the University of California with the construction of the giant cyclotron. It is expected that this machine will be capable of accelerating deuterons to about 60 Mev by next fall. Twice this energy should be obtainable using alpha particles. It is hoped that subsequent improvements will increase the voltage by a considerable factor. Other devices for the same purpose have often been discussed but none of them, to my knowledge, has reached a stage of concrete planning.

Somewhat more advanced is the status of the technique of accelerating electrons to the 100 Mev energy region. For more than a year the General Electric Company has had a betatron of the Kerst-type operating at 100 Mev energy in their laboratories in Schenectady. I visited recently the Schenectady laboratories and saw the machine in operation and could convince myself that the performance is quite smooth and in many ways simpler and more reliable than the operation of a cyclotron. I was also shown results obtained with the machine that I am not at liberty to discuss in this memorandum but which looked to me of very great interest. I am informed that improvements of the betatron are possible which will increase the voltage by a factor between 1.5 and 1.8 without essentially adding to the bulk of the magnet. Very active work is going on at present in order to develop different types of electronic accelerators and some of the proposed schemes look very attractive, although none of them has been tested experimentally so far.

I would consider it very desirable that the Institute should buy from the General Electric Company a betatron identical to the one now in use at Schenectady, and should make at the same time arrangements to have the improvements for raising the voltage from the present value of 100 Mev to 150 or 180 Mev as soon as the technical developments as indicated above are completed. In preliminary talks with representatives of General Electric, the following figures were quoted: cost of the machine of the present type on the assumption that General Electric would receive orders for such machines from five institutions -- $275,000. It appears fairly probable that actually several universities will order betatrons. We know already that MIT and the University of Rochester are both very much interested and other institutions, notably Columbia and California Institute of Technology, have expressed some interest.

Fermi thinks in terms of the highest-energy accelerator available at the time, which was a 100-MeV betatron. The principle of phase stability was discovered by Edwin McMillan and Vladimir Veksler in 1946, and Fermi and colleagues decided to construct a 450-MeV synchrocyclotron, which was completed in 1951.

Walter Bartky   -3-   December 3, 1945

We do not have any figures as to the additional cost of installing the improvements necessary to further raise the voltage of the machine. Firm figures on this point will not be available until the development has reached a more advanced stage. The delivery time for a 100 Mev machine is expected to be of the order of one year. During this time the Institute should take care of preparing a suitable building for housing the machine and devote great attention and effort to the development of experimental equipment to be used in connection with the betatron.

This should include: large Wilson chambers capable of operating with a magnetic field up to 10,000 gauss so as to enable one to measure accurately the curvature of the particle tracks. In a field of this intensity the orbit of a proton of 100 Mev should have a curvature of about 1.5 meters. I believe that one should also try to develop a mass spectrograph for the simultaneous determination of mass and energy of the mesotrons. This apparatus should be planned on an unusually large scale because of the requirements of high luminosity and strong magnetic and electric deflecting fields. In addition one should of course have considerable number of more current devices like counters, ionization chambers, electronic equipment of various kinds, etcetera.

I am convinced that the success of this plan of research will depend to a great extent on the attention and effort put in the development of detecting equipment. If it will be possible to buy the accelerating machine from a commercial firm, the physicists will be in a position to use their time and their specialized skill in a much more efficient way than they could if they were burdened with the task of developing and constructing heavy machinery.

E. Fermi

PARTICIPANTS AT THE INAUGURATION OF THE RESEARCH
INSTITUTES AT THE UNIVERSITY OF CHICAGO, AUGUST 1945

Immediately following the end of the war, the University of Chicago formed the Institute for the Study of Metals and the Institute for Nuclear Studies. These research institutes were formed to attract the extraordinary scientists who had participated in the Manhattan Project. Members of the original staff of the two institutes are shown in figure 5.1.

Figure 5.1 Members of the newly formed research institutes, August 1945. *Seated, left to right:* William H. Zachariasen, Harold C. Urey, Cyril Smith (director of the Institute for the Study of Metals), Fermi, Samuel K. Allison (director of the Institute for Nuclear Studies). *Standing, left to right:* Edward Teller, Thorfin Hogness, Walter Zinn, Clarence Zener, Joseph E. Mayer, Philip W. Schultz, Robert H. Christy, and Carl Eckart. (Photo courtesy Argonne National Laboratory.)

STAFF OF THE INSTITUTE FOR NUCLEAR STUDIES, 1950

The staff of the institute was an extraordinary collection of scientists, most of whom are recognized today. Fermi and Harold Urey were Nobel laureates, and Willard Libby and Maria Mayer became Nobel laureates in 1960 and 1963, respectively. Because of the nepotism rules at the time, Maria Mayer served as a volunteer research associate.

INSTITUTE FOR NUCLEAR STUDIES
STAFF OF THE INSTITUTE

SAMUEL KING ALLISON, PH.D., Director of the Institute for Nuclear Studies and Professor of Physics.

WARREN CHARLES JOHNSON, PH.D., Professor of Chemistry.
WILLIAM HOULDER ZACHARIASEN, PH.D., Professor of Physics.
HAROLD CLAYTON UREY, PH.D., SC.D., Distinguished Service Professor of Chemistry.
ENRICO FERMI, PH.D., SC.D., Charles H. Swift Distinguished Service Professor of Physics.
JOSEPH EDWARD MAYER, PH.D., Professor of Chemistry.
WILLARD FRANK LIBBY, PH.D., Professor of Chemistry.
*EDWARD TELLER, PH.D., Professor of Physics.
MARCEL SCHEIN, PH.D., Professor of Physics.
GREGOR WENTZEL, PH.D., Professor of Physics.
*WALTER HENRY ZINN, PH.D., Professor of Physics.
HERBERT LAWRENCE ANDERSON, PH.D., Professor of Physics.
ANTHONY LEONID TURKEVICH, PH.D., Associate Professor of Chemistry.
HARRISON SCOTT BROWN, PH.D., Associate Professor of Chemistry.
NATHAN SUGARMAN, PH.D., Associate Professor of Chemistry.
JOHN ALEXANDER SIMPSON, PH.D., Associate Professor of Physics.
CLYDE ALLEN HUTCHISON, PH.D., Assistant Professor of Chemistry.
THOMAS HARRISON DAVIES, PH.D., Assistant Professor of Chemistry.
JOHN MARSHALL, JR., PH.D., Assistant Professor of Physics.
MARK G. INGHRAM, PH.D., Assistant Professor of Physics.
JAMES RICHARD ARNOLD, PH.D., Assistant Professor of Chemistry.
RICHARD LAWRENCE GARWIN, PH.D., Instructor in Physics.
GEORGE PARZEN, B.E.E., PH.D., Instructor in Physics.
RAYMOND KAY SHELINE, PH.D., Instructor in Chemistry.
MARIA GOEPPERT MAYER, PH.D., Volunteer Research Associate in Physics (Professor)
GERHART KARL GROETZINGER, PH.D., Research Associate in Physics (Assistant Professor).
DOLORES BANDINI, PH.D., Research Associate in Physics.
LAURENCE BISHOP DEAN, JR., PH.D., Research Associate in Chemistry.
SAMUEL EPSTEIN, PH.D., Research Associate in Chemistry.
LEONA WOODS MARSHALL, PH.D., Research Associate in Physics.
RAMON ORTIZ, PH.D., Research Associate in Physics.
MARTIN STEINBERG, PH.D., Research Associate in Chemistry.
EIZO TAJIMA, A.M., Research Associate in Physics.
JAMES LESLIE TUCK, M.A. (Oxon.), M.SC.), Research Associate in Physics.

CLOVIS ALONZO BORDEAUX, S.M., Junior Physicist.
JOHN M. DORSEY, B.S. (E.E.), Electronic Engineer.
HENRY HINTERBERGER, B.S.M.E., Mechanical Engineer.
LESTER KORNBLITH, JR., S.B.E.E., Chief Electrical Engineer.
LEWIS BEEBE LEDER, S.B., Junior Physicist.
ROBERT HAMILTON LYKKEN, M.E., Mechanical Engineer.
CHARLES RAYMOND MCKINNEY, S.B., Physicist.
BEALY AUSTIN MEADOWS, B.S.M.E., Drafting Supervisor.
RICHARD HENRY MILLER, B.S.E.E., Electrical Engineer.
JAMES BARKER NIDAY, S.B., Chemist.
LEROY SCHWARCZ, S.B., Chief Mechanical Engineer.
WILLIAM SAMUEL SCOTT, S.B., Electrical Engineer.
TOM TANG, S.M., Mechanical Engineer.
ROSE ALTHEA TOMPKINS, A.B., Junior Chemist.
RICHARD A. TRCKA, B.S.C.E., Mechanical Engineer.
AIJI ALVIN UCHIYAMA, A.B., Junior Chemist.
NANCY FARLEY WOOD, A.M., Physicist.

POST-PH.D. FELLOWS, 1949–50

PETER BAERTSCHI, PH.D.
MARCELLO CONVERSI, PH.D.
SHIH-TSUN MA, PH.D.
JOHN LEONARD POWELL, PH.D.

* On leave of absence.

GENESIS OF THEORY OF COSMIC RAY ACCELERATION, 1948–1949

Today, Fermi acceleration is a basic tenet of cosmic ray physics. The original idea was the scattering of cosmic rays off moving magnetic clouds. The calculations can be easily followed. Fermi finds a power law spectrum. The observed power −2.9 is related to the mean free path of scatters, which turns out to be 2 light years, a very reasonable number.

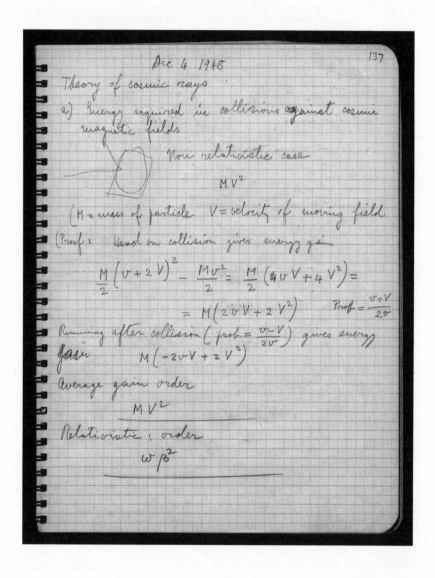

He asks the question, "Why are there no electrons?" He shows that the rate of energy loss by radiation in a typical galactic magnetic field of 10^{-5} gauss is greater than the rate of energy gain for electrons of energy ≥ 250 MeV. He finds this energy to be 3×10^{21} eV for protons! The date of this note is December 4, 1948.

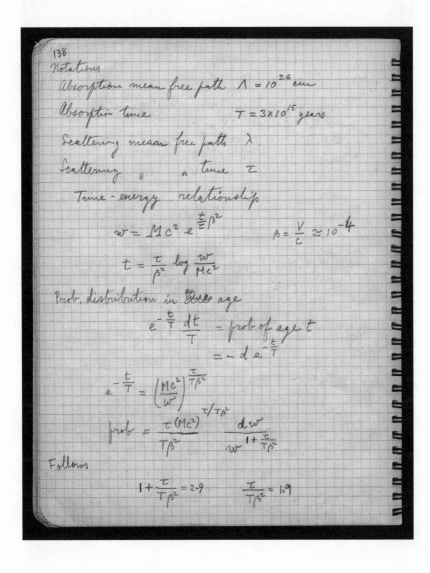

Hence

$$\tau = 1.9\,\beta^2 T = 1.9 \times 10^{-8} \times 3 \times 10^{15} = 6 \times 10^7$$

$$\lambda = 2 \text{ light years!}$$

Why are there no electrons?

Radiation loss is per second

$$\frac{2}{3}\frac{e^2}{c^3} A^2 \frac{1 - \sin^2\varepsilon \frac{v^2}{c^2}}{\left(1 - \frac{v^2}{c^2}\right)^3} = \qquad \text{since for magnetic acceleration}$$

$$\sin\varepsilon = 1$$

$$= \frac{2}{3}\frac{e^2}{c^3}\frac{c^4}{R^2}\frac{w^4}{m^4 c^8} = \qquad HR = \frac{w}{e}$$

$$= \frac{2}{3}\frac{e^2}{e}\frac{e\,w^4}{m^4 c^4}\frac{H^2 e^2}{w^2} = \frac{2}{3}\frac{e^4}{m^4 c^7}H^2 w^2$$

Energy is gained at rate

$$\frac{\beta^2 w}{\tau} \approx 10^{-16} w$$

Hence limiting energy

$$\frac{e^4 H^2}{m^4 c^7} w \approx 10^{-16} \qquad \text{assume } H = 10^{-5}$$

$$w \approx \frac{m^4 c^7}{e^4 H^2} 10^{-16} \approx \left\{ \frac{.6 \times 10^{-108}\ 3 \times 10^{73} \times 10^{-16}}{5 \times 10^{-38} \times 10^{-10}} = 4 \times 10^{-4} \approx 200\ \text{MeV} \right.$$
$$3 \times 10^{21}\ eV \text{ for protons!} \qquad \text{for electron}$$

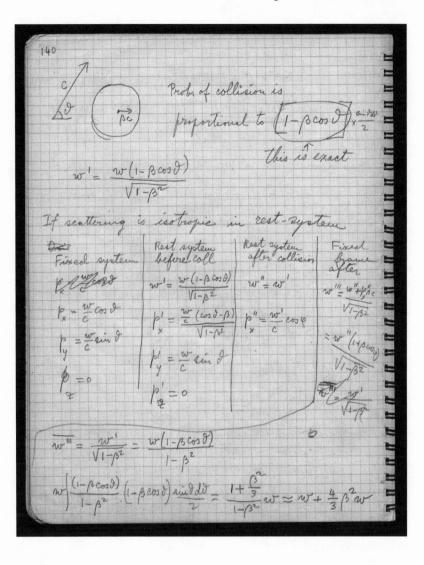

140

Prob. of collision is

proportional to $\boxed{\left(1-\beta\cos\vartheta\right)}\times\frac{\sin d\vartheta}{2}$

this is exact

$$w' = \frac{w\left(1-\beta\cos\vartheta\right)}{\sqrt{1-\beta^2}}$$

If scattering is isotropic in rest-system

Fixed system	Rest system before coll	Rest system after collision	Fixed frame after
$p_x = \frac{w}{c}\cos\vartheta$	$w' = \frac{w\left(1-\beta\cos\vartheta\right)}{\sqrt{1-\beta^2}}$	$w'' = w'$	$w''' = \frac{w'' + p_x'' c}{\sqrt{1-\beta^2}}$
$p_x = \frac{w}{c}\cos\vartheta$	$p_x' = \frac{w}{c}\frac{\left(\cos\vartheta-\beta\right)}{\sqrt{1-\beta^2}}$	$p_x'' = \frac{w'}{c}\cos\varphi$	$= w''\frac{\left(1+\beta\cos\varphi\right)}{\sqrt{1-\beta^2}}$
$p_y = \frac{w}{c}\sin\vartheta$	$p_y' = \frac{w}{c}\sin\vartheta$		
$p_z = 0$	$p_z' = 0$		$\frac{w'}{\sqrt{1-\beta^2}}$

$$\boxed{\overline{w'''} = \frac{w'}{\sqrt{1-\beta^2}} = \frac{w\left(1-\beta\cos\vartheta\right)}{1-\beta^2}}$$

$$w\int\frac{\left(1-\beta\cos\vartheta\right)}{1-\beta^2}\cdot\left(1-\beta\cos\vartheta\right)\frac{\sin\vartheta\,d\vartheta}{2} = \frac{1+\frac{\beta^2}{3}}{1-\beta^2}w \approx w + \frac{4}{3}\beta^2 w$$

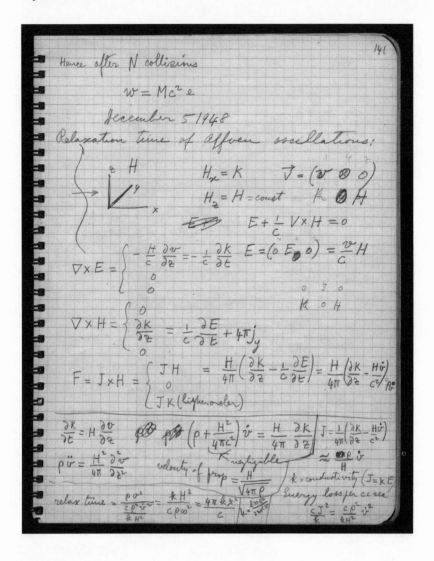

FERMI TO HANNES ALFVÉN, DECEMBER 24, 1948

On December 24, 1948, Fermi wrote a letter to Hannes Alfvén, enclosing a copy of a paper based on the notes of December 4. Fermi confronts Alfvén with a theory of cosmic ray acceleration in contradiction to the ideas of Alfvén and Edward Teller.

The paper was received by the *Physical Review* on January 3, 1945. In the abstract it is remarkable that Fermi points out a difficulty in his theory, in that heavy nuclei are not naturally accommodated. This was because the rate of energy loss due to ionization by nuclei of large atomic number exceeds the rate of energy gain due to the collisions with the moving clouds.

December 24 1948

Dr. H. Alfven
Royal Institute of Technology
Stockholm, Sweden

Dear Dr. Alfven:

I am enclosing the rough draft of a paper on the origin of the cosmic radiation that I propose to send shortly to the Physical Review. You will notice that the ideas that I am presenting are almost directly opposite to those that have recently been discussed by yourself and by Teller.

Of course I am not sure that they are correct but it seems to me that the various orders of magnitude adopted are not unduly stretched and that one can make an apparently good case for my view that the cosmic rays originate and are accelerated primarily in the interstellar space of the galaxy.

I would be very anxious to have your opinion and criticism of this theory. I still remember with great pleasure the most informative discussion that I had with you on your theory of the magneto-elastic phaenomana. As you see I have immediately put to work what I had learned.

With the best wishes for a happy 1949,

                              Sincerely yours

                              Enrico Fermi

ABSTRACT FOR FERMI'S PAPER ON
COSMIC RADIATION, JANUARY 3, 1949

PHYSICAL REVIEW   VOLUME 75, NUMBER 8   APRIL 15, 1949

On the Origin of the Cosmic Radiation

ENRICO FERMI
Institute for Nuclear Studies, University of Chicago, Chicago, Illinois
(Received January 3, 1949)

A theory of the origin of cosmic radiation is proposed according to which cosmic rays are originated and accelerated primarily in the interstellar space of the galaxy by collisions against moving magnetic fields. One of the features of the theory is that it yields naturally an inverse power law for the spectral distribution of the cosmic rays. The chief difficulty is that it fails to explain in a straightforward way the heavy nuclei observed in the primary radiation

COMMENTS BY HERBERT ANDERSON AND
EDWARD TELLER ON FERMI'S PAPER ON COSMIC RAYS

Comments on Fermi's paper on cosmic rays were made by H. L. Anderson and Edward Teller, in *Fermi's Collected Papers*, volume 2, page 665. The accepted theory of shock acceleration by supernova still uses Fermi's idea of the scattering of the cosmic rays by irregularities in the magnetic fields of the shock. This process is known as first-order Fermi acceleration, because the acceleration rate is proportional to the velocity of shock. In Fermi's original theory, the acceleration rate was proportional to the square of the magnetic cloud velocity.

N° 237 AND 238

With paper N° 235 Fermi's experimental work at Argonne came to a close. He gave up his regular excursions to the Argonne for two main reasons. The first was the very stimulating atmosphere which had developed at the University. Many interesting questions came up, and Fermi wanted more and more time to look into them. More and more frequently could he be found at his blackboard examining his equations. The second reason had to do with Mrs. Marshall, who felt compelled to do some of her own work independently of Fermi. These reasons, together with the absence yet of good experimental facilities at the new Institute for Nuclear Studies, combined to convert Fermi once more to theoretical physics.

Paper N° 237 was a direct outcome of heated disputes with Edward Teller on the origin of the cosmic rays. It was written to counter the view that the cosmic rays were principally of solar origin and that they could not extend through all galactic space because of the very large amount of energy which would then be required. Taking up the study of the intergalactic magnetic fields, Fermi was able to find not only a way to account for the presence of the cosmic rays, but also a mechanism for accelerating them to the very high energies observed.

He presented these same views on the origin of cosmic rays, though less extensively, in a talk at the Como International Congress on the Physics of Cosmic Rays (paper N° 238).

H. L. ANDERSON.

N° 237, 238, AND 264

Fermi mentioned to me his interest in the origin of cosmic rays as early as 1946. Several years before that time he mentioned the subject in some lectures in Chicago. He had the suspicion that magnetic fields could accelerate the cosmic particles.

In 1948 Alfvén visited Chicago. He had been interested in electromagnetic phenomena on the cosmic scale for quite some time. At that time I was playing with the idea that cosmic rays might be accelerated in the neighborhood of the sun. I had discussed this question with Alfvén, and he visited us in Chicago in order to carry forward the discussion.

During this visit Fermi learned from Alfvén about the probable existence of greatly extended magnetic fields in our galactic system. Since this field would necessarily be dragged along by the moving and ionized interstellar material, Fermi realized that here was an excellent way to obtain the acceleration mechanism for which he was looking.

As a result he outlined a method of accelerating cosmic ray particles which serves today as a basis for most discussions on the subject. In his papers published in 1949 (N° 237 and 238) he explained most of the observed properties of cosmic rays with one important exception: it follows from his originally proposed mechanism that heavier nuclei will not attain as high velocities as protons do. This is in contradiction with experimental evidence. Fermi returned to this problem in his paper *Galactic Magnetic Fields and the Origin of Cosmic Radiation* (N° 264).

Some details concerning the origin of cosmic rays have not been settled conclusively by Fermi's papers. Another competing theory has been proposed by Sterling Colgate and Montgomery Johnson according to which cosmic rays are produced by shock mechanism in exploding supernovae. The actual origin of cosmic rays continues to remain in doubt.

E. TELLER.

SUMMARY PAGE FROM DATA BOOK ON
MESON-NUCLEON SCATTERING, FEBRUARY 1952

This is a summary page, prepared by Fermi, of many entries by students and colleagues in the logbook. The presentation is a neat and clear summary of results. The logbooks are filled with Fermi's summaries. Discussion about the pion scattering experiments appears in the article by Maurice Glicksman in chapter 8.

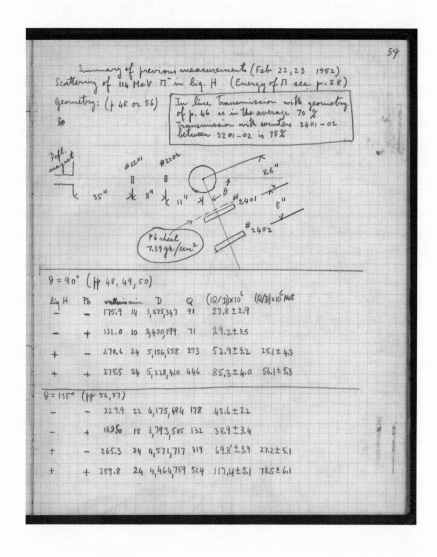

PROGRAM FOR CALCULATION OF CYCLOTRON
ORBITS ON THE MANIAC COMPUTER, 1951

Fermi often did his own work even if it could have been done by a machinist or a student. One can speculate that if there was a problem which lay in the path of his personal research, the most efficient way to solve the problem was to do it himself. Here, he uses the Maniac computer at Los Alamos to calculate the orbits of secondary particles produced on an internal target in the cyclotron. Fermi clearly recognized the power of the digital computer and used it extensively. The computer was used to perform a phase shift analysis of the pion-nucleon scattering experiments, as described by Glicksman in chapter 8. It was also used with John Pasta and Stanisław Ulam to address nonlinear problems with surprising results, as described by Wilczek in chapter 2.

A SUMMARY OF THE EQUATIONS WHICH DESCRIBE THE MOTION OF A CHARGED PARTICLE IN A CYLINDRICALLY SYMMETRIC MAGNETIC FIELD, WRITTEN IN FERMI'S HAND

THE INSTRUCTION SET FOR THE MANIAC COMPUTER

Note that what we call "bit" was then "bigit."

Orders of the Maniac

Logical Symbol	Order Symbol	Address	n(A) = number in A; n(Q) = number in Q; m = number in memory at address.
m → Ac	AA	Address	Write m in A.
m → Ac-	AB	"	Write -m in A.
m → AcM	AE	"	Write |m| in A.
m → Ac - M	AF	"	Write -|m| in A.
m → Ah	BA	"	Add m to A.
m → Ah-	BB	"	Add -m to A.
m → AhM	BE	"	Add |m| to A.
m → Ah - M	BF	"	Add -|m| to A.
m → Q	EB	"	Write m in Q.
X	DA	"	Multiply m by n(Q). First 39 bigits of product appear in A. The 2^{-39} bigit of A is set = 1; Q is cleared.
X'	DB	"	Form m × n(Q). Write first 39 bigits in A, last 39 bigits in Q. Set sign in Q = 0.
÷	DD	"	Divide n(A) ÷ m. Write quotient in Q; Remainder in A.
T	CA	"	Transfer the control to the left hand order of address.
T'	CB	"	Transfer the control to the right hand order of address.
C	CC	"	Perform like T only if n(A) ≥ 0.
C'	CD	"	Perform like T' only if n(A) ≥ 0.
Q → m	EC	"	Write n(Q) into memory at given address.
A → m	DC	"	Write n(A) into memory at given address.
S → m	FA	"	Replace the bigits 8 to 19 of m by the corresponding bigits in A.

Logical Symbol	Order Symbol	Order Address	$n(A)$ = number in A; $n(Q)$ = number in Q; m = number in memory at address.
$S \to m'$	FB	Address	Same for the 28 to 39 bigits.
$HS \to m$	FC	"	Same for the 0 to 19 bigits.
$HS \to m'$	FD	"	Same for the 20 to 39 bigits.
Rn	EE		Replace the contents $\lambda_0 \lambda_1 \ldots \lambda_{39}$ of A and $\sigma_0 \sigma_1 \ldots \sigma_{39}$ of Q by $\lambda_0 \lambda_0 \ldots \lambda_0 \lambda_1 \ldots \lambda_{39-n}$ in A and $\lambda_{39-n+1} \ldots \lambda_{39} \sigma_0 \sigma_1 \ldots \sigma_{39-n}$ in Q.
Ln	DE		Replace contents as above by $\lambda_n \lambda_{n+1} \ldots \lambda_{39} 0 0 \ldots 0$ in A and $\sigma_n \sigma_{n+1} \ldots \sigma_{39} \lambda_0 \lambda_1 \ldots \lambda_{n-1}$ in Q.
$a \to Ac$	EF		Write the 12 bigits of n in positions 0 to 11 and zero in positions 12 to 39 of A.
$a \to Ah$	DF		Add into A the 12 bigits of $\underline{n}$ in positions 0 to 11.
DS	ED	Address	Set the 0 bigit of $n(A)$ to 0.
Print	EA	"	Print m on teletype page printer.
Read	FF	"	Replace m by the next word on teletype tape.
Punch	CF	"	Punch $\underline{m}$ on teletype tape.
Stop	OFF	"	Stop computation.

Instead of address write here the number n.

FERMI'S FLOWCHART FOR A PROGRAM TO CALCULATE THE ORBITS EMANATING FROM A TARGET AT A POSITION $X = r_0, y = 0$

This program is for two-dimensional motion in the median plane of the cyclotron. The initial direction cosines of the orbit $dx/ds = \dot{x}$ and $dy/ds = \dot{y}$ are chosen using the first few terms of the expansion of sine and cosine. The step size is sigma. The program loops until the trajectory exits the fringe field of the cyclotron.

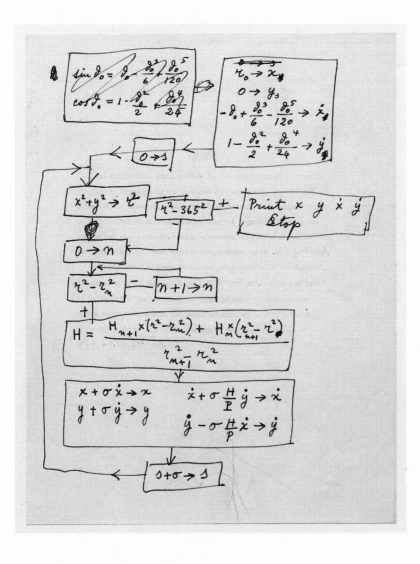

SOME OF FERMI'S NOTES FOR SETTING UP THE CALCULATION OF THE ORBITS

All the data had to be entered as hexadecimal. On one of these pages one can discern that the note had been written on the back of a scrap sheet of paper. It seems that Fermi disliked wasting paper.

$$\ddot{x} = \frac{1}{P} H \dot{y} \qquad \ddot{y} = -\frac{1}{P} H \dot{x}$$

Store into memory

Needed

Table of $H(r^2)$ $0 < r < 365$

40

P, r_0, ϑ_0

3

000 – 0 15 various quantities (variables)

PART OF FERMI'S PROGRAM WHERE HE CALCULATES THE INITIAL DIRECTION COSINES FOR A PARTICULAR ORBIT

QUANTUM MECHANICS EXAM, SPRING QUARTER, 1954

The last course Fermi taught before his death was introductory quantum mechanics, in the winter and spring quarters of 1954. Here is one of the examinations he gave in the spring. Fermi's solutions are given for each problem. The editor of this volume attended this course, as did many others. The auditors outnumbered the registered students.

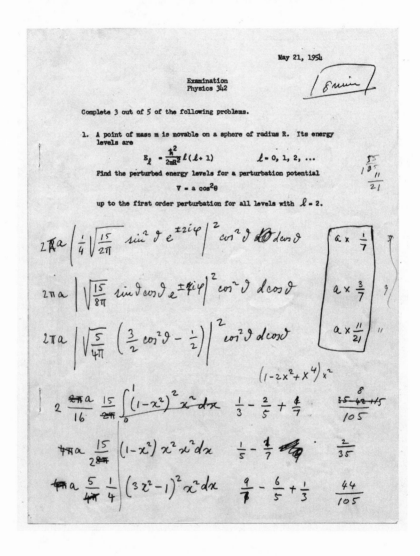

2. A point with electric charge $\underline{e}$ is movable on a straight line and is
    confined on a segment of length $\underline{a}$ by infinitely high potential barriers
    at the two ends of $\underline{a}$. The energy levels are

$$E_n = \frac{\pi^2 \hbar^2}{2ma^2} \, n^2 \qquad n = 1, 2, 3, \ldots$$

Find the selection rules for emission and absorption of radiation.
(Consider electric dipole radiation only.)

n even $\longrightarrow$ n odd

n odd $\longrightarrow$ n even

1 min

3. In applying the Ritz method to the determination of the lowest energy level of the hydrogen atom use the trial function

$$\psi = e^{-kr^2},$$

What is the optimum value of k and the corresponding approximate energy value?

$$E = -\frac{\hbar^2}{2m} \nabla^2 - \frac{e^2}{r}$$

$$\nabla^2 e^{-kr^2} = \left(\frac{d^2}{dr^2} + \frac{2}{r}\frac{d}{dr}\right) e^{-kr^2}$$

$$\frac{d}{dr} e^{-kr^2} = -2kr\, e^{-kr^2}$$

$$\frac{d^2}{dr^2} e^{-kr^2} = \left(4k^2r^2 - 2k\right) e^{-kr^2} \quad \Big|\quad \frac{2}{r}$$

$$\frac{\int\left(-\frac{\hbar^2}{2m}\left(4k^2r^2 - 6k\right) - \frac{e^2}{r}\right) e^{-2kr^2} r^2\, dr}{\int e^{-2kr^2} r^2\, dr}$$

$$I_0 = \frac{\sqrt{\pi}}{2}(2k)^{-\frac{1}{2}} \qquad I_2 = \frac{\sqrt{\pi}}{4}(2k)^{-3/2} \qquad I_4 = \frac{3}{8}\sqrt{\pi}(2k)^{-5/2}$$

$$I_1 = \frac{1}{2}(2k)^{-1}$$

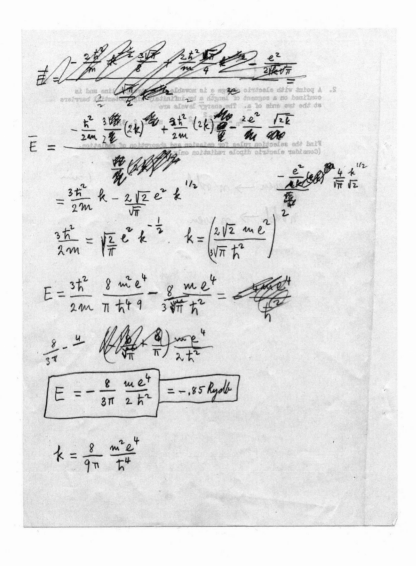

$$\not\equiv \; = \; \cancel{\left[-\frac{2\hbar^2}{m} \; \frac{k}{} \; \frac{3}{2} \; \frac{\sqrt{\pi}}{2} \; + \; \frac{2\hbar^2}{m} \; k \; \frac{}{} \; \frac{}{4} \; \frac{}{} \; - \; \frac{e^2}{2k\sqrt{\pi}} \right]}$$

$$\bar{E} \; = \; \frac{-\dfrac{\hbar^2}{2m} \, \dfrac{3}{2} \dfrac{\sqrt{\pi}}{2} (2k) \; + \; \dfrac{3\hbar^2}{2m}(2k) \; - \; \dfrac{2e^2}{m} \dfrac{\sqrt{2k}}{}}{\cancel{}}$$

$$= \frac{3\hbar^2}{2m} k - 2\frac{\sqrt{2}}{\sqrt{\pi}} e^2 k^{1/2} \qquad \xleftarrow{\qquad} \qquad -\frac{e^2}{k\sqrt{\pi}} \frac{4}{\sqrt{\pi}} \frac{k^{1/2}}{\sqrt{2}}$$

$$\frac{3\hbar^2}{2m} = \sqrt{\frac{2}{\pi}} \, e^2 k^{-\frac{1}{2}} \qquad k = \left(\frac{2\sqrt{2} \, m e^2}{3\sqrt{\pi} \, \hbar^2} \right)^2$$

$$E = \frac{3\hbar^2}{2m} \frac{8 \, m^2 e^4}{\pi \, \hbar^4 \, 9} - \frac{8}{3} \frac{m e^4}{\sqrt{\pi} \, \hbar^2} = \cancel{\frac{4 m e^4}{\hbar}}$$

$$\frac{8}{3\pi} - \frac{4}{} \qquad \left(\frac{8}{\sqrt{\pi}} + \frac{8}{\pi} \right) \frac{m e^4}{2 \hbar^2}$$

$$\boxed{E = -\frac{8}{3\pi} \frac{m e^4}{2 \hbar^2} \; = -.85 \, Rydb}$$

$$k = \frac{8}{9\pi} \frac{m^2 e^4}{\hbar^4}$$

4. A point of mass m is scattered by two equal potential holes *A and B* as in the figure. Assume the potential holes of depth V and radius R with R ≪ ƛ (de Broglie wave length). Using the Born approximation find the differential scattering cross section for particles of momentum p traveling parallel to the x-axis.

p o o

$p\cos\vartheta$ — —

$$\frac{d\sigma}{d\omega} = \frac{p^2/v^2}{4\pi^2\hbar^4}\left|\int U\,e^{\frac{i(\vec{p}-\vec{p}')\cdot\vec{r}}{\hbar}}\right|^2$$

$$\frac{4\pi}{3}VR^3\left\{e^{i\frac{pa}{\hbar}(1-\cos\vartheta)} - e^{-i\frac{pa}{\hbar}(1-\cos\vartheta)}\right\}$$

$$\frac{8\pi i}{3}VR^3\sin\left\{\frac{pa}{\hbar}(1-\cos\vartheta)\right\}$$

$$\frac{p^2/v^2}{4\pi^2\hbar^4}\left(\frac{8\pi}{3}VR^3\right)^2\sin^2\left\{\frac{pa}{\hbar}(1-\cos\vartheta)\right\}$$

5. An atom in a state of angular momentum 3/2 is initially in a state with $J_x = 3/2$. Its angular momentum state is subsequently analyzed as to the possible values of its component J_z along z. What are the probabilities of the 4 possible values of J_z?

$$J_x = \frac{1}{2} \begin{vmatrix} 0 & \sqrt{3} & 0 & 0 \\ \sqrt{3} & 0 & 2 & 0 \\ 0 & 2 & 0 & \sqrt{3} \\ 0 & 0 & \sqrt{3} & 0 \end{vmatrix}$$

$$\begin{pmatrix} 0 & \sqrt{3} & 0 & 0 \\ \sqrt{3} & 0 & 2 & 0 \\ 0 & 2 & 0 & \sqrt{3} \\ 0 & 0 & \sqrt{3} & 0 \end{pmatrix} \begin{vmatrix} x \\ y \\ z \\ t \end{vmatrix} = \begin{vmatrix} 3x \\ 3y \\ 3z \\ 3t \end{vmatrix}$$

$\sqrt{3}\, y = 3x$ $\left| \begin{array}{l} t = x = 1 \end{array} \right.$

$\sqrt{3}\, x + 2z = 3y$ $z = y = \sqrt{3}$

$2y + \sqrt{3}\, z = 3z$ $2z = 3\sqrt{3} - \sqrt{3} = 2\sqrt{3}$

$\sqrt{3}\, z = 3t$ Normalized results

J_z	probabilities
3/2	1/8
1/2	3/8
−1/2	3/8
−3/2	1/8

OUTLINE FOR THE SPEECH "THE FUTURE OF NUCLEAR PHYSICS," UNIVERSITY OF ROCHESTER, JANUARY 10, 1952

"The Future of Nuclear Physics" was delivered at the University of Rochester on January 10, 1952. Here the notes are handwritten. There is a reference to the practical aspects of nuclear energy. Among the practical aspects, Fermi always stressed the medical applications. Then he warms up to his current interest, which is cosmic rays and particle physics. There is the planned humor of comparing the designation of the particles to a list of Greek letter fraternities. Most interesting is the conclusion, for which Fermi runs out of space and begins to write with a smaller and smaller hand, finally finishing in a bubble of empty space above. The text of this conclusion, which we have typeset in full size, is an eloquent argument for the value of basic research. It also shows Fermi's often expressed concern for the future of mankind.

Some of you may ask what is the good of working so hard merely to collect a few facts which will bring no pleasure except to a few long hairs who love to collect such things and will be of no use to anybody, because only few specialists at best will be able to understand them? In answer to such question I may venture a fairly safe prediction. History of science and technology has constantly taught us that scientific advances in basic understanding have sooner or later led to technical and industrial applications that have revolutionized our way of life. It seems to me improbable that this effort to get at the structure of matter should be an exception to the rule—What is less certain, and what we fervently hope is that man will soon grow sufficiently adult to make good use of the powers that he acquires over nature.

Chapter Six

Reminiscences of Fermi's Faculty and Research Colleagues, 1945–1954

❖ ❖ ❖

Richard L. Garwin
WORKING WITH FERMI AT CHICAGO
AND POSTWAR LOS ALAMOS

I entered graduate school at the University of Chicago in the fall of 1947. After some months of getting accustomed to the course work, I felt the need to do some experimental work, which was my interest and my strength. So I summoned my courage and went to Professor Fermi to volunteer to help in his laboratory. I discovered that he had his own small machine shop with a lathe, band saw, and other tools. He had great respect for the ability of the central shops in Chicago, but he felt they were too fastidious—a job submitted to the central shop was returned with ten times the precision necessary, which often took ten times longer than necessary. Fermi made a lot of his own equipment and this was just my milieu.

I encountered in his laboratory Leona Woods Marshall—Fermi and Leona had just done some work on electron-neutron interaction—and Jack Steinberger, who was working away in the corner there with Geiger tubes and piles of graphite and lead doing something with cosmic rays. Fermi and Marshall were doing an experiment on positronium using Geiger tubes; they had built the tubes themselves incorporating a cotton thread soaked with sodium-22, a radioactive material which would emit a positron. They were studying the annihilation process. But Martin Deutsch at MIT was studying positrons as well, and he soon scooped them because he had access to scintillation counters with RCA photomultiplier tubes. Eventually Fermi got such phototubes, and we decided to incorporate them in future experiments.

I found that the electronics being used at that time was rather primitive. It consisted of Rossi coincidence circuits from the 1930s which had microsecond resolving time and Geiger tubes with a hundred microsecond dead time. It was time to make an investment in building equipment that could be used more efficiently for the future. I took it upon myself to develop new electronics. I first made some fast pulsers, so I could test the circuits I was to build. I invented some coincidence circuits with resolving times of a few nanoseconds to take advantage of the fast pulses coming from the photomultpliers. The coincidence circuit used current switch technology and a diode clamp. They were used for quite awhile at Chicago and adopted elsewhere.

Fermi had a technician upon whom he was quite dependent. Nevertheless when the young technician had the opportunity to advance his career in another position Fermi encouraged him to accept it. As a consequence, Fermi had to train a new technician—my first understanding of mundane limitations on even the greatest.

Fermi's propensity for building his own equipment was shown when the cyclotron was put into operation. It was a synchrocyclotron. There were a lot of the problems with high-power radio frequencies; whenever one builds a device which must provide an electric field over a large area, one finds that all of the power leaks out in an unsuspected place and something burns up. This is even more of a problem when the frequency is swept from a high frequency down to a lower frequency—as is required by the mass of the proton increasing from 938 MeV at rest to 1388 MeV (the proton rest energy plus a kinetic energy of 450 MeV). This RF power leak was just one of many technical problems that Fermi helped solve.

Fermi would meet every morning at 7:30 with the engineers, headed by Leroy Schwartz. He would get a report on the problems and prospects of the cyclotron activity; Fermi would talk to the engineers about the ideas and analyses he had made of the progress of the previous day and then go about his work and hear from them either later that evening or the next morning.

When the cyclotron started to work, obviously there were beams that had to be brought out and targets that needed to be positioned within the cyclotron. So rather than having a large number of targets mounted on probes penetrating the vacuum chamber that could be plunged into position, Fermi decided that he would have a movable target—a trolley that would move along the rim of the cyclotron. He took advantage of all the natural properties of the cyclotron—it has a large magnetic field which can serve as the stator of an electric motor. Fermi built a cart that moved along the ridged pole in a circumferential manner. The cart had a little target that could be flipped in and out of the beam. After the trolley was made, he decided to put a thermocouple on the target so that one would know how much power was being dissipated in the target by the nuclear interactions. I remember about 35 watts was considered a good performance for the cyclotron. I helped him a little bit there. Although Fermi could control the cyclotron very well, less capable people were bothered by the time lag between changing the beam intensity incident on the target and the time the target would take to come up to temperature as measured by the thermocouple. So I designed a little circuit which anticipated what the result would be. The time lag for adjusting the intensity was reduced from 90 seconds to only a few seconds.

Fermi suggested at lunch that one could use an analog computer for Robert Mulliken's molecular calculations. He was also interested in nuclear shell structure in the behavior of nuclear particles in very peculiar nuclear potentials—there were Feenberg models and there were Jastrow models of the nuclear potential, among others. One day, I came into Fermi's lab in Ryerson Hall. He had a coil (or maybe it was, in this case, a bar magnet) suspended as a torsion pendulum. He was going to make an analog computer. The analog of the spatial distribution of a one-dimensional wave function was the time behavior of this torsion pendulum, and the analog of the difference between the energy and the depth of the potential would be the current fed to the coil as a restoring force. It was a perfect analogy—for small angles, anyhow.

If we were going to carry this out we needed to record the position of the coil as a function of time and program the current representing the restoring force as a function of time. I suggested that this electromechanical technique was probably not the best way to solve the problem. One could eliminate all the mechanical parts and build the analog computer just using operational amplifiers. Having made the suggestion, I was obligated to build it. In those days it involved a whole rack of equipment with my specially designed operational amplifiers swinging between plus and minus 600 volts. Fermi used the system a bit, and after his death I believe Clyde Hutchison used it, but no one else.

Among my major failures was to not follow one of Fermi's suggestions. In December 1949, I received my Ph.D. and joined the faculty of the physics department. I had my own laboratory in the institute building. I was busy doing experiments on the betatron; planning experiments for the cyclotron; getting ideas where I could—including scintillation counter ideas from Gaurang Yodh and others. Fermi came in one day, and he gave me some suggestions to do a theoretical calculation of nuclear wave functions and energy levels. I gave the idea some thought but took no action. Two weeks later he came back and asked what progress there had been, and I told him none. So he said he was going to talk to Maria Mayer about it. I just simply lacked the courage to put down what I was doing and take up a new field, where in fact I would probably not have done nearly so well as Maria in creating the field of nuclear shell structure.

Now, in those days, Chicago paid faculty nine-month salaries. One could either starve or get government contracts to provide salaries for the other three months, but I wasn't about to do either. Fermi suggested that I could be a consultant to Los Alamos. Rumor has it, although I don't recall, that I had made some suggestions to him about nuclear weapons. Fermi said that the place to work on such things was Los Alamos. So in 1950, with my wife Lois and our son Jeffrey I went to Los Alamos for three months. Fermi and his wife Laura were there. I shared an office with Enrico Fermi, which was a good experience. People would come in and talk to him, and I was stimulated by the discussions. During the war Fermi had gone to Los Alamos, not when the laboratory was formed, but later, because he had been busy with the planning for the plutonium production reactor at Hanford. He went to Los Alamos in the fall of 1944 and remained through 1945. He was not in charge of any development group, although there was a Fermi group. He was a treasured consultant also known as the pope. Whenever people needed some knowledge which was lacking or had problems which they could not solve themselves, they would seek Fermi's help. One way or another he would either show them how or, in extremis, provide an answer.

In 1950 Fred Reines came into our office and suggested that, with all of the nuclear explosions going off at the test site, one could put a detector immediately underground and detect the neutrinos. In response to the suggestion Fermi pointed out that a nuclear reactor—one of the modern reactors—burns several kilograms of uranium each day, while a nuclear weapon of 17 kilotons requires the fission of only one kilogram of uranium. In any case, the neutrinos come from beta decay of the fission products, and not from the fissions themselves. Thus, many more neutrinos are available from a reactor, and one can place a detector closer to the neutrino source. I believe that Fermi's suggestion to Fred Reines led Clyde Cowan and Reines to a more feasible and ultimately successful experiment at a reactor.

Stan Ulam would come in to the office, and he and Fermi would work together on calculations for the burning of a cylinder of deuterium. This was the classical idea for the hydrogen bomb ("super"), which, I believe, had been Fermi's original suggestion to Edward Teller in 1941. The idea had caught fire with Teller but not in reality, as it turned out to be very difficult calculation. I will not go into the details. Fermi would take an accountant's blank spreadsheet, a Marchant calculator, and a slide rule, and with these tools he would convert the partial differential equations that were involved to the first-order differential equations. He would fill in the first few rows of the spreadsheet, each row being an interval of time. Fermi and Stan would talk about the parameters. Then their computer, Miriam Caldwell, would arrive, take the spreadsheet, and return the next morning with the results. Fermi and Ulam would plot the results and then give her the next problem.

Part of my work that summer was to begin experiments to measure with more precision the deuterium-deuterium and deuterium-tritium cross sections down to incident energies of 15–20 kilovolts. Previous measurements of these cross sections at the University of Texas dated from before the war. I thought this was pretty weak evidence to rely on when deciding to build hydrogen bombs as the cross sections were poorly known. So I devised an experiment and began to build it. When I had to leave at the end of the summer to return to my responsibilities at Chicago, Fermi encouraged Jerry Kellogg and the laboratory director, Norris Bradbury, to import Jim Tuck from England to continue the experiment. Tuck had been at Los Alamos during the war. He and his team completed the experiments, which were published in 1994.

I also worked on diagnostics of nuclear explosions. The first couple of weeks that I was at Los Alamos in 1950 I spent in the classified library reading the weekly reports of all of the groups written during the war and after. When I returned for the summer in 1951, I didn't share an office with Fermi. The Physics Division decided I would be better off as a consultant to the Theoretical Division, and that is where I was ever after in my summers at Los Alamos.

Fermi was concerned that summer with a large number of things, including Taylor instabilities. If one takes an upright glass of water, its surface is stable. Ripples and waves are generated on the surface when it is perturbed. If one places a card over a full glass of water and turns it over, the water remains in the glass: it is stable. Once the card is removed the interface is still in equilibrium supported by the air pressure, but soon the water falls out, as the equilibrium is unstable. This instability is a very important phenomenon in nuclear weaponry and had been plaguing the people at Los Alamos ever since they considered implosion weapons in 1944. Actually, these instabilities were considered in 1942, but they really had to be understood in order to use plutonium in 1944.

So in 1951 Fermi had schematized the problem on his blackboard. Everybody knows that, in the beginning stages of Taylor instability, you assume a ripple on the surface, and instead of behaving sinusoidally in time, it behaves exponentially in time with the same time constant, except it is real instead of imaginary. So there is a time in which the amplitude doubles; the next interval it quadruples; the next interval it gets to be eight times as big. And pretty soon, of course, this cannot go on, because the energy in the instability exceeds the energy that was driving it; the velocity exceeds the velocity of light. And so the question is what happens at large amplitudes?

Fermi said, "Let me make a model; I'll have a broad tongue which moves into the dense material; I'll have a narrow tongue that moves away from it and I'll just solve this numerically." Fermi made set up the analysis, but he wasn't quite satisfied. One afternoon around 4:50 p.m. John von Neumann came by and saw what Fermi had on the blackboard and asked what he was doing. Enrico told him, and von Neumann said, "That's very interesting." He came back about fifteen minutes later and gave him the answer. Fermi leaned against his doorpost and told me, "You know, that man makes me feel I know no mathematics at all."

There was another time at Chicago when our colleague, Edward Teller, who was also on the faculty (although you couldn't tell it by the amount of time he spent here) came by and told Fermi of his most recent enthusiasm. And as Teller left, Fermi commented, "That's the one monomaniac I know with more than one mania."

I left the University of Chicago in December 1952 for a new IBM laboratory in New York to change my focus from particle physics to condensed matter physics. I didn't like having to tell people six weeks in advance what I wanted to do with the cyclotron or to work with a team of six people. Now, of course, it's sixty weeks in advance and six hundred people, so I think I made the right choice considering my personality. But I certainly admire what has been done in particle physics since. At IBM I studied superconductors, and liquid and solid helium, and I continued to work summers at Los Alamos.

By 1954 I had worked for a year (sort of half-time) on air defense of the

United States and had made contact with people in Washington outside the nuclear weapons community. Upon hearing of Fermi's illness I returned, I think, in October, and saw him in his house. He clearly had an inoperable cancer. And we talked. He regretted not having been more concerned about public policy. I was, of course, terribly saddened that Enrico Fermi was taken from us at the age of 53. I just imagine the joy that he would have experienced had he been around to see the evolution of computers and the development of physics, and the role that he might have played if he had been given another twenty or thirty years.

❖ ❖ ❖

Murray Gell-Mann
NO SHORTAGE OF MEMORIES

It is really difficult for me to pick out a particular set of memories of Enrico. As I think back to those years (1952, 1953, and 1954), almost half a century ago, I experience a flood of recollections of so many different kinds.

I recall the lunches at the round table at the Quadrangle Club, with Enrico and other colleagues like Harold Urey, Bob Mulliken, John Platt, Willy Zachariasen, Bob Gomer, and so many others. Enrico would talk with delight about the latest *Li'l Abner* comic strips—what other European immigrant professor would even have heard of *Li'l Abner*? One day, reminded of the period in Rome when he was called Il Papa (and Amaldi was called Il Cardinale, George Placzek was called Il Santo, and so forth), Enrico explained how he could still become pope in reality if he joined the Catholic hierarchy right away. It would take him so many years to go from the priesthood to the rank of archbishop, so many more to become a cardinal, and then, depending on the longevity of the Holy Father, so many years to succeed him. (In all of this, he was ignoring any obstacle presented by his marriage to Laura!)

He would gently poke fun at Mulliken's calculations of molecular properties by asking whether, instead of performing digital computations, one might not employ an analog computing device, and if so what better device than the molecules themselves?

I have vivid memories of the Monday afternoon theoretical meetings involving Enrico, Gregor Wentzel, Murph Goldberger, Ed Adams, and me, and occasionally one or two others. Those meetings were held in Wentzel's office, because Enrico couldn't stand having his office filled with the smoke from Gregor's cigars. Many times I would ask a question and Enrico would say, "Well, let us see," go to the blackboard, and answer it, typically starting at the upper left and finishing at the lower right with the correct answer in a box. He loved to do that, but what he liked a lot less was facing up to things he didn't fully understand.

Sometimes those things were received ideas that he was right not to understand because they weren't really correct. One was the usual treatment of quantum mechanics. Why, he would ask, isn't the moon all spread out in its orbit? Is it really because astronomers are watching it? I didn't believe, either, that the Copenhagen interpretation was fundamental, and I was delighted that Enrico was put off by it as well.

Fermi was also impatient with the claims that the beta-decay interaction (the so-called Fermi interaction or four-fermion interaction) had been shown to have a particular form. The prevailing story was that the interaction was scalar and tensor rather than the vector or axial vector that appeared in Enrico's original paper (where he actually had a vector current times an axial vector current). Of course, we learned a few years later that the claims of S and T were in fact based on wrong experiments and that V and A were correct.

He told us about the lessons on computers that he was getting at Los Alamos from Nick Metropolis. He also described the work he was carrying on there with John Pasta and Stan Ulam on a model of a nonlinear string with many equal weights attached, equally spaced. Enrico had hoped to see how energy, if concentrated initially in one normal mode of the linear approximation, would gradually distribute itself among many such modes as the nonlinearities produced increasing disorder. Instead, the computations showed that nearly all the energy flowed back into the first mode, and then a somewhat smaller but still very large fraction came back again into the first mode, and so on. Nowadays, that behavior is less mysterious than it was then.

Sometimes Enrico could be stubbornly wrong, as when he kept blasting away at the idea that his pion-nucleon scattering experiments should be interpreted in terms of a $J = 3/2$, $I = 3/2$ resonance. He harped on a phase shift solution he had found in which there was no resonance. Finally, Maurice Glicksman had to consult Hans Bethe and bring back an alternative solution in which the phase shift did exhibit resonant behavior by increasing through 90 degrees. Even then, Enrico argued against the resonance. Of course, in reality it is there. Moreover (and this would really have horrified him), it is even a good idea to interpret the resonance as an unstable particle, especially since it turned out to have a metastable partner, the omega minus.

Naturally I discussed physics with Enrico at other times as well, not just on Monday afternoons. I have written about his reception of my ideas about strangeness. He discouraged me by saying that, like Richard Feynman, he was more convinced than ever that the new particles were states of high angular momentum, with decay slowed down by the centrifugal barrier. That night, I wandered around the institute, disconsolate, ending up in the office of our theoretical secretary, Vivian, who had succeeded Gabrielle Falk. Violating rules of gentlemanly behavior, I looked at a letter she was typing for Enrico. It was to Giuseppe Cocconi, thanking him for his calculations on the high angular momentum model but warning him that a young theorist named Gell-Mann was speculating about a very different idea that might explain the new particles. On seeing that letter, I became much less discouraged, but quite angry with Enrico.

We all knew how Fermi would often raise an objection at a seminar talk and then not allow the seminar to proceed unless the speaker dealt satisfactorily with his objection. If the speaker couldn't satisfy him, the seminar was effectively over. Naturally, I was somewhat apprehensive when I came to teach a class on fields and particles and found that Fermi was going to attend every session.

My previous class (my first one ever) was on mathematical physics, and it was really terrible. I frequently made up problems with no solutions, and some of the bright students, such as Jim Cronin and Horace Taft, would gently point out the flaws in my assignments.

I did better in the class on particles and fields. Although the presence of Enrico in the back row was terrifying, nothing went seriously wrong. One day, in the course of explaining the strangeness idea, I discussed the neutral mesons K^0 and $\bar{K}^0$, which are antiparticles of each other. Fermi objected that nothing new was involved, since one could take, instead of the K^0 and $\bar{K}^0$ fields, their sum and difference, which would then describe two neutral spin zero particles, each its own antiparticle.

Fortunately, I was prepared for this remark. The year before, I had had great difficulty getting my letter on strangeness published by the *Physical Review*. For one thing, I had to change "strange" or "curious" particles to "new unstable" particles. More important, the referees challenged the idea of a neutral spinless particle that was distinct from its antiparticle. I looked up the prewar paper by Nick Kemmer (whom I had met in Cambridge in 1952) proposing an isotopic triplet of mesons, with a single neutral member and two oppositely charged members. In describing the neutral meson, he started with the charged spinless particles of Wolfgang Pauli and Victor Weisskopf and took away their electric charge, leaving precisely two neutral mesons that were antiparticles of each other. Then he had to show that he could take the sum and the difference of the two fields, throw one of them away, and get a single neutral particle that was its own antiparticle. So what needed to be justified in 1952–53 was easy to accept in the late 1930s, and vice versa. When I supplied that reference to the *Physical Review,* the editors gave up.

Thus, I was able to answer Enrico by saying that in certain decays of the neutral Ks, the sum and the difference would indeed be the relevant fields, but that in the production process a single K^0 could accompany a lambda hyperon, while a single $\bar{K}^0$ could not. Later, when I wrote up the neutral K particle situation, I thanked Fermi for his question.

So many personal memories come back. Skiing, for me, meant going to Alta, Utah, with Bob Gomer, and I loved Bob's jokes about people who talked of skiing the headwall at Wausau, Wisconsin. But one morning Enrico came to collect me to go skiing on a small hill near Chicago. When we got there, I was appalled, seeing nothing but a gully filled with soil and rocks and a few flakes of snow. But Fermi would not be deterred by anything. He put on his skis and tackled the slope. What a wimp he must have thought me to be, since I just watched him, fascinated but not eager to follow.

He liked to tease young people, saying he didn't understand their attitudes. That was hard on his children sometimes, as I remember his daughter Nella telling me. And to politically engaged students like Art Rosenfeld, he would say, for example, "I do not understand what you young people have against the atomic bomb."

When Enrico lay dying in Billings Hospital, I realized how much I cared for this brilliant, funny, difficult man. I was on leave in the East, and I invited Frank Yang to come with me to Chicago to see him. When we got to the bedside, Enrico kept telling us not to be downcast. "It is not so bad," he said. He told of a Catholic priest who had visited him and whom he had had to comfort. And Frank reminded me a few years ago of what Enrico said when we left, never to see him again. "Now, it is up to you."

❖ ❖ ❖

Marvin L. Goldberger
ENRICO FERMI (1901–1954): THE COMPLETE PHYSICIST

Fermi was a physicist, first and foremost. In the words of his long-time colleague Gilberto Bernardini, a colorful mangler of the English language, "Fermi was a physicist with a capital F." That's what he was, all the time—

First published in *Physics in Perspective* 1 (1999): 328–336.

a physics machine, with little time for or interest in anything else. He accepted the challenge of Nature and was prepared to use any tool at his disposal to meet that challenge. There was a kind of dogged single mindedness to him that once prompted Eugene Wigner, a Nobel Laureate and one of the towering figures of twentieth-century physics, to remark, "I wish I could think about one problem at a time as Fermi does." Wigner was known for snide remarks cloaked in superficially complimentary forms and this was one of them, although he did admire Fermi. But Fermi could not do otherwise; once he got his teeth into a problem he would not let go.

I first met him in January of 1946 shortly after he came to Chicago, before I got out of the Army and entered graduate school. Fermi had, of course, been in Chicago earlier to work on the Manhattan Project when he built the world's first man-made nuclear reactor, the reactor that reached critical mass for a chain reaction on December 2, 1942, and opened a new era in human history. Obviously Fermi did not do it all by himself, but this fateful experiment was his baby and his triumph. In the course of my work in the theoretical physics section of the Chicago project I read many of Fermi's reports. They were typically short, lucid and surprisingly easy to follow.

There were Project stories told about the omniscient Fermi. For example, there were many important neutron cross-sections that had not been measured but were needed for the design of the Hanford plutonium-production reactor—which was the Chicago Project's assignment. To determine the estimates of these cross-sections for calculations, the "Principle of Minimal Fermi Reaction" was used. It worked as follows: You suggested a value of the cross-section so high or so low that Fermi became very agitated. When you reached the point where Fermi shrugged indifferently, you wrote down the number.

After the war, Fermi came to Chicago with a whole group of young and middle-aged stars in tow, many of whom he had helped to recruit. Among them were Harold Urey, Joe and Maria Mayer, Edward Teller, Herbert Anderson, Robert Christy, John and Leona Marshall, Eldred Nelson, Stanley Frankel and Nick Metropolis. There were also a flock of students who, like me, had undergraduate degrees in physics and had worked on the Manhattan Project during the war—most of them at Los Alamos. Fermi was the Pied Piper who brought them to Chicago. I was one of the exceptions; Teller, who had arrived in late 1945, recruited me. C. N. "Frank" Yang, a brilliant Chinese student, was another who came on his own initiative after he found that Wigner, whose work and style Yang greatly admired, would be unavailable at Princeton and that Fermi was no longer at Columbia. Frank (he and some of his Chinese fellow students decided that they should take on American names which they chose purely on the basis of sound and not because they had any relationship to their real first names) was awarded the Nobel Prize along with another Chinese Chicago student, T. D. Lee, in 1957. An unusual group of graduate students made up that first postwar class.

By virtue of our war work we were much more mature in physics than the usual holders of the B.S. degree. Fermi played a central role in our graduate education. He not only taught regularly but also met once or twice a week with a special subset of the graduate students, mostly those he had brought from Los Alamos, and discussed a series of topics—some of his own choosing, others suggested by the students. Topics ranged from astrophysics to statistical mechanics, from Bose-Einstein condensation to general relativity. In most cases he had detailed notes kept in magnificently cross-indexed notebooks to refresh his memory. For a long time I was not invited to become a member of this privileged group and I was terribly envious of those who were. Eventually, after I became Fermi's student, I made it, and it was a real treat. Fermi's obvious pleasure in sharing his deep insights was inspiring.

These discussions showed us what it means to be a real physicist. You work everything out, you never delude yourself by saying that you probably could do something if you really wanted to. You do it and you write it down so that you can recover it when necessary. Fermi was religious about this. His notebooks were incredibly clear—he would work things out on the blackboard, get them absolutely right and then copy the essentials into those carefully indexed notebooks. He once flirted with the idea of taking Polaroid pictures of his blackboard to save time and effort, but he really didn't find that satisfactory. On countless occasions I tried to follow his example, but I never had the self-discipline necessary to index and cross-index my own notebooks. The basic lesson I, like all of Fermi's students learned was described by Frank Yang this way: "Physics is to be built from the ground up, brick by brick, layer by layer."

Another of Fermi's work patterns was the development of what he called his Magic Memory or sometimes his Mechanical Memory. This was a collection of formulas for things Fermi felt he would need frequently in doing research, like the radiation from a charge moving at relativistic speeds, quantum-mechanical sum rules, vector-addition coefficients, solutions of the Dirac equation in a coulomb field, hydrodynamic phenomena, heat transfer, etc. (Non-experts need not be concerned about this jargon.) Only useful eternal truths made their way into this single-file drawer reserved for the Mechanical Memory and Fermi never questioned the validity of anything in there. If it was there at all, it was correct and ready to be used. Fermi is reported to have said that he could solve only nine problems in physics. His great gift was to be able to force everything he ever encountered into one of nine molds—all enshrined in the Magic Memory, I presume.

Fermi's regular courses I found strange. They were carefully organized and interesting, but to me at least somewhat unsatisfactory. He would illustrate general principles with the simplest examples. Nothing was ever very hard. If there was a difficulty, Fermi usually had invented at some time, perhaps in connection with an entirely different work, a trick to get the answer. But the unfortunate student confronted with a hard problem of a similar variety couldn't invent for himself a corresponding trick to save the day.

Fermi was also prepared to lie outrageously for pedagogic purposes. Occasionally I had to give his lectures when he was traveling. He always had the lectures written out in great detail but I often found I could not make myself say the words because I knew they were wrong. He felt that in certain cases you should simply not insist on presenting the absolute truth. I guess my basic complaint was that his courses didn't prepare you for the real world, where, if you don't happen to be Enrico Fermi, not all problems are easy.

Fermi's greatest influence as a teacher was the example he set and through his style of doing research. He did not have a large number of students compared with some other well-known physicists. He often complained that he had never developed a "school," as Robert Oppenheimer (the charismatic leader of the atomic bomb development at Los Alamos) had before the war. He was not so impressed with Oppenheimer the scientist, but he was openly envious of Oppenheimer the teacher, whose worshipful entourage of students trailed behind him from Berkeley to Caltech and back; students who walked, talked, and lit women's cigarettes just as their guru did. Nonetheless, all of us who had the privilege of being Fermi's students were worshipful in our own way.

Fermi the person was more complex than Fermi the scientist. He enjoyed social interactions as long as they didn't take up too much time, but he was always reticent about his feelings and was certainly not the person you would go to with a serious personal problem. He was not cold, but not warm either.

One of Fermi's greatest pleasures was interacting with students and colleagues at lunch at the Commons (the main campus dining hall) at the University of Chicago. There he loved to discuss how to solve problems of various kinds. One problem I remember was to estimate the number of railroad cars in the U.S. He would follow to a logical conclusion the consequences of a few simple assumptions, and come out with an answer that was always correct to within a factor of two. I often suspected that he had worked these exercises out during his regular periods of insomnia commencing at 4:00 a.m. Surprisingly, Fermi was very bad at puzzles, even very elementary ones, that require a single flash of insight. He could never resist the challenge, but he seldom got the right answer. I took great comfort from that—I mostly couldn't do them either.

He enjoyed square dancing, and such dance parties were regularly held at his house. He also prided himself on his physical stamina and would shame his students into extended exposure to a brutally cold Lake Michigan. He played tennis without form but with a relentless tenacity; he was a human backboard. When asked why he didn't learn proper strokes, he would say: "Is it not the purpose of this game to win? Is it not true that I almost always win?" Form without substance was not his style either in tennis or in physics.

He regarded Nature as an adversary that he could win against in a similar way. He had a disdain for formalism in physics even though he had considerable formal power. But when he used it, for example in his famous 1930 paper on quantum electrodynamics, it was always aimed at getting at real problems. If he encountered a complicated integral in his work, he rarely fooled around with analytic approximations; rather, he went directly to his desk calculator. Those of us who emulated his style sometimes missed general principles in our zeal to get answers. A specific example: I remember once in the course of a calculation involving sums over products of vector-addition coefficients the answers came out startlingly simple. I was intrigued by this, but in my rush to get some cross-section calculated I missed an opportunity to independently discover Racah coefficients although, in retrospect, it's far from clear that I could have done so anyway.

Fermi also felt that time spent agonizing over the interpretation of quantum mechanics was a waste. He had some discomfort with the topic but not the profound Einsteinian philosophical misgivings. He was disturbed, for example, that in classical physics, given the forces acting on a particle and two pieces of information (the initial position and velocity), one could calculate the trajectory; by contrast, in quantum theory the value of the wave function at all points in space at time zero, an infinite number of values, had to be specified. But once having expressed his uneasiness, he would set about merrily doing quantum-mechanical calculations and comparing them with experiment.

Some abstract theoretical issues did worry him. For example, the various so-called conservation laws (such as conservation of energy, momentum and angular momentum) that follow from invariance principles. Invariance principles require that a description of an isolated physical system should not depend on our choice of an origin of time, or whether we translate the system in a particular direction in space or capriciously decide to rotate the physical system before making observations. He was acutely uncomfortable with another of these general invariance requirements; namely, that there could be no experimental distinction between a system and its mirror image, that Nature had no "handedness." He used to say over and over, "Who turns his glove inside out?" In 1952, three distinguished theoretical physicists, Eugene Wigner, Gian-Carlo Wick and Arthur Wightman wrote a profound paper on certain rules in quantum theory that follow from these invariance principles that we discussed at great length in Chicago. There is a famous footnote in that paper that comments on the absence of experimental support for parity conservation, the consequence of the mirror image invariance, in the weak interactions. In the same footnote, the authors point out the possibility that neither parity nor charge conjugation (another of these invariance ideas relating to relations between particles and their anti-particles such as electrons and positrons, distinguishable only by their opposite charge) is conserved but only their product. This intrigued Fermi.

It is always hard to understand how people get ideas. Fermi once told me what led him to develop what are known as Fermi-Dirac statistics and the Fermi theory of beta decay. As for the former, he said that he had had an unbelievably hard time trying to understand Wolfgang Pauli's fundamental paper on the exclusion principle. (Anyone who has tried to read it can appreciate the problem.) But, Fermi said, on the very day that he had finally mastered it he invented the statistics. My guess is that his immediate aim was to apply this newly mastered idea to a different problem in order to convince himself that he really understood it. (I'll come back to this point in another connection shortly.)

Although Fermi was totally comfortable in 1934 with both the concept and formalism associated with creation and annihilation operators in connection with the quantization of the electromagnetic field, he had not absorbed the corresponding formalism for matter and in particular for spin one-half fields. In connection with beta decay, there were arguments that made it impossible to accept the idea that the emitted electrons and neutrinos were somehow really inside the radioactive nuclei tunneling out like alpha particles. Thus they had to be created in the decay process. Pascual

Jordan and Wigner had developed the requisite machinery in a classic paper on representations of field operators that satisfied anti-commutation rather than commutation relations. The paper was quite abstruse and emphasized mathematics rather than physics. Fermi recognized, however, that he needed something like this and finally mastered the formalism. Again, he told me, on that very day he wrote down and worked out the beta-decay theory. Any physicist who has not read Fermi's 1934 paper on beta decay should rush out and do so immediately. In my opinion it is the very epitome of what a scientific paper should be. The problem is stated clearly, a solution is presented, and the results compared with experiment. No smooth talk, no pretension, no promise that this is the first of a long series, etc. Just the facts! It should be required reading for every physics student.

Of course, even Fermi didn't invent gunpowder every time he sat down at his desk. But he was always prepared. He was a perpetual student who would, for example, decide to learn group theory and having made such a decision, he would teach himself the subject with minimal reference to books. The way he read theoretical papers was to look at the abstract, close the journal, work it out for himself and compare his result with the author's. For an experimental paper he would try to extract the raw data and reduce and interpret it for himself. The pattern of understanding some idea and applying it to a new situation was a characteristic one. My own thesis grew out of a need on Fermi's part to understand the Monte Carlo method developed at Los Alamos to deal with neutron diffusion and the dynamics of nuclear explosions. He suggested I look into the passage of a high-energy neutron through a heavy nucleus by following in detail the fate of each of the partners of a collision with a target nucleon using probabilistic methods to calculate the location and consequences of the initial and subsequent collisions.

Fermi had an uncanny ability to analyze a complex system by building a model that could be solved exactly and that reproduced the essential features of the system. Many physicists can write down an exact and hopelessly complicated description of some phenomenon that is totally intractable and yields no insight. To do what Fermi did is infinitely harder and not easy to teach, especially the necessity of recognizing the domain of validity of the simplifications. (I often worry that the existence of supercomputers will discourage young physicists from trying the Fermi approach. That will be a disaster.)

Fermi had another amazing talent. He could concentrate his full powers on a problem someone else was interested in, no matter how far afield it might be from his own interests. A weekly informal seminar, held at the

Institute for Nuclear Studies, was a superb showcase for this talent. There, chemists, biologists and physicists of all stripes would give talks. Fermi took enormous pleasure in these events and taught the speakers a great deal about their subject. Edward Teller occasionally spoke on various topics, and a frequent Fermi preamble to a remark was "What Edward is trying to say is. . . ." This would then be followed by an extraordinarily lucid presentation of what Teller might have said had he really understood the subject.

There was another forum involving only theoretical physicists. It began in 1948 following the Pocono Conference on quantum electrodynamics. Fermi wanted to understand the new formulation of the theory as given by Julian Schwinger in a marathon sixteen-hour presentation at that conference. So Fermi, Gregor Wentzel, and a few of his graduate students— Frank Yang, Geoff Chew, Jack Steinberger, and I—met regularly for several weeks studying this work. I still have the notes.

When I returned to join the faculty at Chicago in 1950, we began a weekly lunch with theoretical faculty only, usually held in Wentzel's office over sandwiches. There we talked about on-going work and new papers, but mostly it was an opportunity for us to have a go at Fermi. I remember at one of those lunches asking him about the adiabatic theorem in quantum mechanics. He said, "I once thought about that; let me see if I can remember." He went to the blackboard and delivered a brilliant thirty-minute lecture on the subject. Afterward I found a paper he had written twenty years earlier, and amazingly, it was almost line for line the same presentation he had just given.

Fermi's talent for concentrating on your problem made it a pleasure to be his colleague. Having done his own work between 4 a.m. and 8 a.m., he was fully prepared to spend the day solving everyone else's problems. You could almost invariably count on getting help on some aspect of your own research, some insight you hadn't thought of. In addition, Fermi was not a nay-sayer. If you went to him with an idea, even a half-baked one, his first instinct was to see if it could be made to work. I think my nearly pathological aversion to people whose pleasure in life consists of looking for destructive counter-examples comes from my association with Fermi. This positive thinking approach was very important for me when Murray Gell-Mann and I were struggling with the birth pains of dispersion relations. We were strongly encouraged by Fermi to pursue this work even if we could not give very convincing derivations. That, he said, could wait for later.

Fermi was also a gifted experimental physicist. He felt that for himself there were times to be a theorist and times to be an experimentalist and his instincts for this timing were superb. Al Wattenberg, a Fermi experimental student, in an article for the *European Journal of Physics* asserts that Fermi was an "excellent and very careful experimentalist." I have no reason to doubt this. I do know that he took enormous pleasure in working with his hands in the shop at Chicago. He once remarked that being an experimentalist gave one a great advantage over theorists: "Whenever you don't have an idea you can go to the lathe and make something." During one of his experimental phases he decided that he wanted to understand electronics more deeply, not to depend so much on the people in the shop to design apparatus. In his characteristic way, he posed himself the problem of building some kind of circuit. He then wrote down 27 non-linear differential equations that described the system. Appalled by the complexity, he decided that relying on the shop wasn't so bad after all.

It was in connection with both experiment and theory that Fermi got interested in computers. He had, of course, seen the need for computation in bomb design during the war and had been a close associate of John von Neumann and Stanislaw Ulam during and after the war. In an effort to master the computer, he, John Pasta and Stan Ulam studied a "toy" problem of a set of equal-mass particles connected by non-linear springs. They started the system out in one of the normal modes of linearly coupled masses and expected to find an equipartition of energy among all the others. Indeed for short times this happened, but then to their surprise nearly all the energy came back into the original mode. It was not until the pioneering work of Martin Kruskal and his collaborators that this behavior could be understood in terms of the initial normal-mode splitting into solitons and the recurrence phenomenon analyzed precisely. Another thing behind this investigation of Fermi's was his uneasiness about ergodic behavior of systems and alleged proofs of ergodicity.

Inevitably Fermi had some involvement in policy issues. He was a reluctant participant in the affairs of the American Physical Society (APS). When in the natural course of events it was suggested that he run for president of the APS he at first refused, but someone told him he simply must do it. He was, of course, elected. In those days the president was a rather minor figure because the Society was managed, and I mean managed, by K. K. Darrow, the executive secretary. Darrow found, however, that Fermi was no patsy. On an occasion when Fermi wanted to call an unscheduled meeting of the executive committee, he had to remind an outraged Darrow which of them was in fact the president.

As for the national-security issues that many of the Manhattan Project warriors had become involved with, Fermi was again a largely reluctant participant. He did, however, accept membership on the General Advisory Committee of the Atomic Energy Commission chaired by Robert Oppenheimer. This was at a time when the Committee wrestled with the profound issue of whether the U.S. should embark on a crash program to develop thermonuclear weapons, H-bombs. The urgency of the issue was the result of the explosion of a fission bomb by the Soviet Union in the spring of 1949, well in advance of the time General Groves and various politicians had expected. The Committee unanimously recommended against the development of H-bombs on both technical and moral grounds, a recommendation that later played an important role in the crucifixion of Oppenheimer in 1954. In an addendum to the main report, Fermi and I. I. Rabi made a very powerful statement: "It is clear that the use of such a thermonuclear weapon cannot be justified on any ethical ground which gives a human being a certain individuality and dignity even if he happens to be a resident of an enemy country." They go on to say: "The fact that no limit exists to the destructiveness of this weapon makes its very existence and the knowledge of its construction a danger to humanity as a whole." In spite of these obviously deep-seated reservations about the H-bomb, Fermi did pursue some theoretical ideas connected with it in collaboration with Stan Ulam in the summer of 1950. President Truman had made the decision to push ahead with a crash program and Fermi was not the sort to take to the streets to oppose a perceived national need. In addition, it was a fascinating physics problem and in that summer work he and Ulam showed that the then-current design ideas would not work. When called upon to testify in the Robert Oppenheimer hearing, Fermi repeated his great misgivings about the decision to proceed with the H-bomb and his feeling that the world would have been a better place had that particular child been stillborn. His testimony during the Oppenheimer hearings focused entirely on the deliberations of the General Advisory Committee and had no significant impact in the hearings. On his deathbed Fermi told Richard Garwin that he felt in retrospect that he should have tried to play a greater role in policy issues.

Fermi died tragically young, at 53, on November 29, 1954. To those who knew him he was a constant joy to be around. His vitality, his enthusiasm for physics and life were contagious. Those of us fortunate enough to have been his students were unbelievably lucky. His death was as dignified as his life. He willed himself to live through a meeting of the Physical Society in Chicago where many of his friends came to say goodbye. The evening the meeting ended he passed away. We shall not see the likes of him soon, if ever.

The Comman Man

George Gamow relates the following story illustrating how simply Enrico Fermi met a problem with instant results:

Fermi was a sturdy Roman boy with a great sense of humor. While he was still a professor in the University of Rome, Mussolini awarded him a title: 'Eccellenza' (His Excellency). Once he had to attend a meeting of the Academy of Sciences at the Palazzo di Venezia, which was strongly guarded because Mussolini himself was to address it. All other members arrived in large foreign-made limousines driven by uniformed chauffers, while Fermi drew up in his little Fiat. At the gate of the Palazzo he was stopped by two *carabinieri* who crossed their weapons in front of his little car and asked his business there. According to the story he told to the author of this book, he hesitated to say to the guards: "I am His Excellency Enrico Fermi," for fear that they would not believe him. Thus, to avoid embarrassment, he said: "I am the driver of His Excellency, Signore Enrico Fermi." "*Ebbene*," said the guards, "drive in, park, and wait for your master."

George Gamow, *Thirty Years that Shook Physics: The Story of Quantum Theory* (Garden City, N.Y.: Doubleday, 1966), pp. 140–141.

❖ ❖ ❖

Roger Hildebrand
FERMI'S CLASSROOMS

In 1947 a fellow student, Herbert York, hailed me one day on the Berkeley campus, saying I should go with him to a classroom in Le Conte Hall to hear Enrico Fermi talk.

I had heard nothing about it. I had thought of Fermi as someone you read about in textbooks, not as a professor who might drop into a classroom unannounced to talk to anyone like me. Well, he had *not* come to talk to anyone like me. He said that most of us in that packed room must be there by mistake. He had expected a seminar with just a few theoretical colleagues interested in the technical aspects of a specialized topic. He said that he would not be offended in the least if all but the specialists left the room. When no one budged he shrugged and said that the best he could do was to give a five-minute introduction that he hoped would satisfy most of us, and that he would then pause while we went off to more important tasks.

He gave the five-minute introduction and then urged any of us to leave who did not find the topic rewarding. Surely, he said, that would be most of us. Again no one budged. Then he turned to face the blackboard and said that since he could no longer see us no one should feel embarrassed to sneak out.

There was no way to sneak out but I bolted for the door, and I was sent off with a burst of laughter from the audience. I went home, opened an elementary text, and then heard a knock at my door. It was Herbert York. Fermi had paused again and that time Herb was the only one to leave.

The next time I heard Fermi talk—it was in 1952 in a classroom here, in Ryerson Laboratory—I understood, or thought I understood, every word. I had in the meantime had four years of graduate physics at Berkeley, but more to the point, Fermi was talking this time to his own class. When he explained something you hadn't thought about he made it so clear that you whispered, "Of course! Why couldn't I have thought of that myself?"

I was not one of the students. I was there because Fermi had asked me to take his class the following week when he would be away. He asked me because I had just arrived and had no regular teaching duties in my first quarter at Chicago. He ended his lecture saying, "Next Tuesday, Professor Hildebrand will give you a superb lecture on the deuteron." Fearing that the students might not agree, I asked him, after class, if there would be any time before that Tuesday when I could go over the subject with him. He said, "Surely. Would 7:30 tomorrow morning be convenient?" That is when he usually arrived at his office.

As you can imagine, I was there at 7:30, deeply conscious of being the newest, greenest member of the faculty. I went to his blackboard, picked up the chalk, and began to say and write what I knew about the deuteron, responding from time to time to his very simple questions. Somehow it all came out better, much better, than it had when I rehearsed the night before. And that is how it was when you talked with Fermi. If an idea came up that had never before crossed your mind, you enjoyed, for the moment, the illusion that you *had* thought of it yourself.

Then, there was a classroom for the faculty. We met on Thursday afternoons, in room 480 of the Research Institutes Building, for a sort of jam session. The meeting was never announced. No one came with notes. There was no script.

Gregor Wentzel, sitting in a big leather chair at the front left, served as unofficial MC. He would sometimes start the discussion by pointing at one of us in the rear, saying, "Well Murph (or Val, or Dick, or Murray), what are you thinking about?" You couldn't say you weren't prepared: You had to be thinking about something. But usually someone was eager to start: not to talk for an hour, but to tell of an idea, an observation, or a just a rumor, to be picked up, examined, embellished, explained, or destroyed as the discussion progressed. I'll try to give examples that capture the breadth and spirit of the thing.

John Simpson told about cosmic rays from the sun; Gene Parker, about how those cosmic rays got from sun to earth; Clyde Hutchison, about nuclear magnetic resonance (what hospitals use nowadays to make expensive scans of your innards); Harold Urey, about synthesis of amino acids in a replica of the earth's first atmosphere; Dick Garwin, about an experiment to produce in the cyclotron what was then called a V-particle; and Murray Gell-Mann, facing a barrage of questions from Gregor Wentzel, about a quantum number that would not be conserved if a V-particle appeared alone. That exchange left most of us bewildered. But, as was often the case, there was a sequel. On another Thursday, Wentzel said, "This is how I understand what Murray was saying," and his account of Murray's idea was lucid and convincing.

Most often Fermi was the star. He told of nucleon resonances, cosmic ray acceleration, and galactic fields. But what he really enjoyed was solving someone else's puzzle.

A visitor had measured the thickness of a superfluid film on a vertical surface at points above the liquid bath. Fermi said, "Don't tell us the answer. Let's see if we can work it out." And he did.

Someone proposed an experiment that was clearly a long shot, arguing that the potential value of the unexpected result justified the effort. Fermi said immediately that the point to consider was not the value of a certain outcome, but rather the value of an experiment to determine the outcome. He showed in three lines how the value *you* assign to an experiment should depend on your own bias about the outcome. This was just a flourish, a finger exercise that was over in a minute. Sam Allison asked him if he had *ever* computed the bias-dependent value of his own experiments. Fermi smiled and said, "Never."

A final example: Willard Libby told us of a technique to date samples of water. (Alas, it was a technique ruined by H-Bomb explosions.) He could tell how long ago the last rainwater had seeped into a deep cave or when the grapes were harvested to make your glass of wine. (When you visited his lab you sometimes enjoyed vintage aromas). Now he wanted to know whether the age of water at the bottom of the ocean could be explained by waves churning the surface. Again Fermi said, "Let's work it out." And with obvious satisfaction, he worked the problem through, introducing characteristics of monster waves that he had learned on a transatlantic voyage (it's the big ones that count).

This was not a seminar about physics, or chemistry, or astronomy; certainly not about high-energy physics, or nuclear chemistry, or extragalactic astronomy. It was about science as pursued for the love of it by ardent participants. It was something that could flourish only in the midst of abundant, eclectic virtuosity. And while it did flourish, room 480 was the best of classrooms.

❖ ❖ ❖

Darragh Nagle

WITH FERMI AT COLUMBIA, CHICAGO, AND LOS ALAMOS

The story begins for me back in 1940 at Columbia University when, as a shiny new graduate student, I thought I should pay a courtesy call on the illustrious Professor Enrico Fermi, whose presence at Columbia was one reason for my coming. His office was listed as seventh floor of Pupin physics laboratory. So I took the elevator and got off at that floor. There was a sort of foyer serving two wings. I looked around for the office. I wondered whether he would be very formal, as I had heard some Europeans were, and would require an appointment set up with his secretary. Suddenly . . . a door burst open, and a dark-haired man ran past me at full tilt, disappearing through the opposite door. He was wearing a lab coat and was carrying a bit of something in a pair of chemical tongs. I stood there flat-footed wondering what to do, when the first door burst open again, and what appeared to be a younger version of the first gentleman, similarly attired and similarly burdened, dashed by me and also disappeared through the second door. Although the full significance of this minidrama was not completely clear to me then, it seemed clear enough that this was not the day for a courtesy call on the professor. The bit of something that was being carried was probably a rhodium foil, which had been activated by neutrons. Because rhodium decays with a half-life of 42 seconds, one had to be quick in carrying it from the lab where it was activated to a place where its activity could be measured. Only later did I find myself caught up in races devised by Fermi and Herbert Anderson.

Soon afterward I was to meet Fermi in the classroom. The course was mechanics, elasticity, hydrodynamics, and a short introduction to quantum mechanics. The course was exceptionally clear, beautifully organized. He wrote everything out on the board: as a consequence I have complete notes of the course. He also gave a course in thermodynamics, using his book. Other faculty members—Maria Mayer, Arnold Nordseick—also gave great lectures. Edward Teller, on the other hand, was interesting and informative, but quite disorganized.

In February 1942 Fermi transferred his activities to the University of Chicago. It was December of that year that he directed the historic startup of the pile in the west stands of Stagg Field. I always thought of the pile as a sleeping, malevolent monster, and the Gothic style entrance to the stands as appropriate for some kind of horror movie (fig. 6.1). Inside you saw demonic figures, with red eyes glaring in faces black as pitch (although of course it was graphite) scurrying about. Instead of chains there were the control rods to restrain the monster (fig. 6.2). Outside the students and citizens of Chicago plodded through the snow, unaware.

A few months later the pile was torn down and moved to the Argonne Palos Park site. Later, after the pile had been rebuilt and brought back into operation, a vigorous program of experimentation was instituted, and Fermi would often be at Palos Park. Sometimes he would stay overnight in the dorm. At night there was no food service, so we did our own thing in the kitchen. Fermi would sometimes make a frittata. I remember his sitting up

Figure 6.1 The west stands of Stagg Field. Under these stands the first controlled nuclear chain reaction took place on December 2, 1942. (Photo courtesy Argonne National Laboratory.)

Figure 6.2 An exposed layer of the pile during assembly. In the holes of the graphite blocks spheres of natural uranium oxide are placed. (Photo courtesy Argonne National Laboratory.)

in bed one night, reaching for a little notebook, writing a few lines, and then going back to sleep.

Later, many from the "Met lab" moved to Los Alamos. Anderson and I drove out together. We were met by Fermi, who characteristically took us on a hike: to the Upper Crossing of Frijoles Canyon, in the course of which he told us about some current activities.

Fermi had a little group headed by Herb, which carried out experiments at Omega site. Perc King was in charge of the Water Boiler, the enriched reactor we used for measurements of cross sections, etc. Others in the group were myself, Julius Tabin, Joan Hinton, Bob Carter, and Jim Bridge. We shared the building with a group doing critical assembly work: Fermi was deeply suspicious and ordered us to stay away whenever critical assembly tests were planned. We would go up to the mesa or sometimes take a hike, often with Fermi. Sadly, he was proven right by the tragic, needless deaths of Louis Slotin and Harry Daghlian. I have often wondered why Fermi did not intervene with Robert Oppenheimer.

Following the Trinity test in July 1945, one wanted to collect soil samples from the crater a few hours after the explosion. The samples were to be analyzed by the radiochemists to obtain the ratio of the fission products to the unreacted fissionable material. From this one got a value for the yield of the explosion. The Anderson-Nagle-Tabin method was to use a Sherman tank to go into the crater, to dig up samples thru a hole in the bottom of the tank. The tank bottom and personnel compartments were lined with lead (fig. 6.3).

Julius Tabin, Anderson, and I took turns going into the crater, and of course there was a driver. It was dangerous because if the tank stalled, there

Figure 6.3 Military tank used to collect soil samples following the Trinity test, July 1945. *Left,* Darragh Nagle; *right,* unidentified driver. (Photo courtesy Los Alamos National Laboratory.)

was no escape; we would have cooked. Our luck held, we got our samples, and the tank didn't stall, but we all got significant doses of radiation.

Fermi's method was simple, practical, and safer. He had George Weil fit out a second Sherman tank with rockets that were to be fired into the crater and retrieved with cables attached to the rockets. As it turned out, the second tank stalled some distance away from where it was supposed to be, and the cables got scrambled up. The experiment failed, but nobody got hurt.

I joined the faculty of the University of Chicago in 1952, with the intention of working on the new cyclotron. Herb Anderson told me how the cyclotron happened. After the war, scientists were held in the highest regard in the United States, Fermi in particular. The story is that Urner Liddell, chief of the nuclear physics branch of the Office of Naval Research, came to Chicago and asked, "What instrument would Fermi like to have?" Herb said to Fermi, "What would you like? I'll build it. An accelerator, a big computer, . . . You name it." Fermi chose a cyclotron. The ONR was as good as its word providing funds (along with the Atomic Energy Commission), and also making available the facilities of the New York Naval Shipyard for much of the construction. The machine followed principles developed at Berkeley. Indeed they provided drawings and invaluable advice. But the success of the Chicago project was due in large measure to the engineering and organizing skills of Herbert Anderson.

In 1952, the cyclotron was at the stage of final assembly and testing in the new building known as John Marshall's Barn. A view of the cyclotron is shown in figure 6.4. I had come back to Chicago, and one day I had taken time off from hunting leaks to sit in my lab and read some reports. Suddenly, Herb burst through the door in a manner reminiscent of our first encounter years ago. He demanded, rather truculently I thought, to know what I was doing. I said politely that I was thinking of building a liquid hydrogen target in order to study pion-proton scattering. He said, politely this time, that that was a good idea, and to get going. So I got out of my chair and went across the street to the famous west stands to see Earl Long of the Institute for the Study of Metals, who then was very helpful to me in getting a hydrogen target built quickly out of simple materials.

Anderson, Fermi, and I began to measure the total cross sections of hydrogen at several energies, first for negative pions, then for the charge-exchange reaction, followed by deuterium and hydrogen cross sections for several energies of negative and positive pions. Later we were joined by Anderson's students Ron Martin, Maurice Glicksman, and Guarang Yodh, who made important contributions and added to the pleasure of the work.

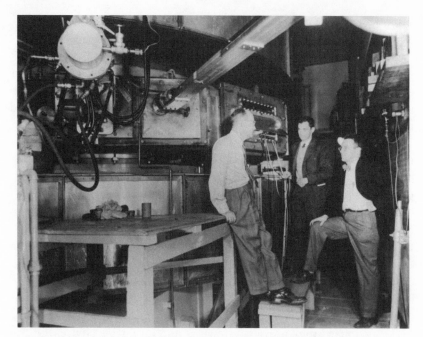

Figure 6.4 View of the Chicago synchrocyclotron. *Standing, left to right:* Fermi, Herbert Anderson, and John Marshall. (Photo courtesy University of Chicago visual archives.)

The results showed the cross section for negative pions rising about linearly with the energy up to about 100 MeV and then appearing to level off at about 60 millibarns, the "geometrical" value, that is, pi times the pion Compton wavelength squared. At the time this type of behavior seemed not unreasonable. The positive pion cross sections for a more limited range of energies seemed to be larger, rising, but not yet "leveling off."

At first we were puzzled: why were the positive pion cross sections larger, considering that only one reaction channel was open for *pi plus,* whereas for *pi minus* three channels were open: namely, elastic scattering, charge exchange, and radiative capture?

One day Anderson and Fermi were in the cyclotron control room working on the experiment. Fermi was operating the counters, and Anderson was reading his mail. Herb said, "Here's a letter from Keith Brueckner. He says he can explain our cross sections." Fermi grunted something like "What does he know?" Anderson said, "It says here that the sections should be in ratios 9:2:1, and there should be a peak around 180 MeV." At this point Fermi grabbed the paper, went up to his office, and left Anderson to run the counters. Fermi reappeared soon and explained it all to him.

Briefly, Brueckner introduced a phenomenology based on ideas from strong coupling meson theory, and the charge independence of nuclear forces, including the pions. He predicted a resonance at about 180 MeV for *pi plus*, and that near 180 MeV the cross sections for the *pi plus* elastic reaction, the charge exchange reaction, and the *pi minus* elastic scattering would be 9:2:1. Strong suggestions of this seemed to be in our data.

We then set about measuring the angular dependence of these reactions. The target was modified to accommodate this. Partial wave analysis of the data was done first by Fermi, then by the rest of us, and by many others. In particular, Fermi and Nick Metropolis, working at Los Alamos with the Maniac computer, did analyses requiring energy continuity of the several partial wave amplitudes. The Maniac found several plausible solutions, one of which showed the $P_{3/2}$, I-spin 3/2 resonance. It took a long time to convince everyone (including Fermi) that this latter one was the correct one. The work of Martin and Glicksman was important in this connection.

Fermi was proud of his strength and stamina, and liked to lead on hikes, swims, and so on. During the west stands times, we would swim in Lake Michigan, Fermi in the lead, and Herb, me, Tabin, and Leona Woods following along. He liked tennis. He liked to ski: in New Mexico we would ski at places like Sawyer's Hill, the bowl above Camp May, and Hyde Park.

I remember a story about Fermi and Emilio Segrè. His old friend was trying to persuade Fermi that trout fishing was a great sport. Fermi was not persuaded.

Segrè: "But you don't understand, Enrico. It requires great technique: you must select just the right fly, cast it, and make it move through the water just so, and then the fish will think it's a real fly. . . ."

Fermi: "Ah, I see, a battle of wits!"

❖ ❖ ❖

Valentine Telegdi
REMINISCENCES OF ENRICO FERMI

My first encounter with Fermi occurred during the 1947 Basel/Como international conference. It turned out to be much more a meeting (albeit a virtual one) with Dick Garwin than with Fermi: I had asked Fermi about his graduate students and what they were working on. Fermi mentioned Garwin's work on β-γ coincidences, done with the then-new scintillation detectors. By chance I had a colleague at Eidgenössische Technische Hochschule Zürich (ETH) who was working on the same topic. By comparing his and Garwin's approach, I realized Garwin's superiority, a conclusion that survived fifty years.

When I arrived, in June 1951, in Chicago, I was not stunned by the senior faculty. I expected them to be capable of anything—if Fermi had walked on water, I would not have been astonished! But I was overwhelmed by my untenured colleagues, in particular Garwin and Murray Gell-Mann, both my juniors by seven years and both omniscient. My friends, it was a time when you could be proud to be the dumbest one!

The junior physics faculty was full of a spirit of fun. That spirit manifested itself particularly during the yearly Christmas parties, where students and faculty competed in all sorts of tasks. One year we erected a large box with flashing lights, representing a computer and clearly labeled ENRIAC—it was supposed to provide order-of-magnitude estimates in reply to any question put to it, thus mimicking Enrico. Then we told the plot: Gell-Mann had invented the "world formula" (which made all future research needless) and his senior colleagues had killed him, placing his corpse in the cyclotron's vacuum tank. We asked the ENRIAC, "There is a body in the cyclotron tank. What should we do?" The machine answered (in Enrico's voice, imitated by me, inside the box): "The average male weighs 65 kg. Forty percent of that mass is water. The speed of our vacuum pumps is x liters per minute. The body is well desiccated by now and there is no point in opening up the tank." It was also revealed that Gell-Mann had shown his formula to a graduate student, so that his demise had been useless. Fermi laughed most of all.

Fermi's greatness has been described most succinctly by his Italian colleague G. Bernardini: "That man, that man, is what I call a Physicist with a capital F."

What did make Fermi so special?

As a *physicist*, his universality, his versatility (both in experiment and in theory), and his constant disposition and ability to share his knowledge.

As a *person*, his modesty, his lack of neuroses, and his man-on-the-street mannerism.

As an *Italian*, his frugal lifestyle, his lack of interest in food and drink, elegant clothing, and titles.

Fermi gave you the illusion that his way of thinking was just like yours—only clearer, sharper, and better oiled. No jumps, no flashes, no strokes of genius. The genius lay in the result. In his presence we all became a little smarter.

Let me conclude by telling a few anecdotes that may illustrate his character and his sense of humor and gift for irony:

Fermi exercised his gift for providing order-of-magnitude estimates sometimes under surprising circumstances. When William Zachariasen, a distinguished colleague, was in the hospital recovering from a heart attack, Fermi decided to pay him a visit there. Zachariasen complained bitterly that he was given too little to eat, only 1500 calories per day. Fermi asked him, "Will, you are a great reader of detective stories, are you not?" Zachariasen replied, "Yes, I am." Fermi then asked, "Willy, how long does it take a corpse to cool?" to which Zachariasen replied, "Four to five hours." After some thought about heat losses, Fermi concluded, "Then you cannot possibly survive on 1500 calories!"

At some point, Fermi decided to teach his private seminar Group Theory. He took out his index cards on that subject and started to discuss first Abelian groups, then Burnside's theorem, and only much later the concept of a group itself. Some of the students were confused by this unorthodox approach. The master said, "Group theory is merely a compilation of definitions." Therefore he had simply followed the *index* at the end of Weyl's book.

Fermi looked at his surroundings mostly with a physicist's eyes. Once, answering a question of Bill Libby's (in the institute seminar) about mixing in the ocean, he devised instantly an equation describing that phenomenon. There was only one single parameter in it, λ, the wavelength of surface waves. For this, Fermi inserted a numerical value of 200 meters. Somebody in the audience asked, "Enrico, is it not rather 600 meters?" to which Fermi replied, "Maybe so. But it was certainly 200 meters when I last crossed the Atlantic." During a trip to Brazil, one of the things that impressed him most was "that the moon increases there on the opposite side from what it does here."

As a member of the Royal Italian Academy, Fermi carried the title "Excellency," abbreviated as S.E. (meaning His Excellency). Fermi said, "I am confused. I do not know whether I am '*My* Excellency' or '*His* Excellency'!"

Somebody once asked Fermi, "What do Nobel laureates have in common?" Fermi replied, "Not much. Not even intelligence."

❖ ❖ ❖

Al Wattenberg

FERMI AS MY CHAUFFEUR (FERMI AT ARGONNE NATIONAL LABORATORY AND CHICAGO, 1946–1948)

In 1946, at the end of the war, Fermi returned to Chicago. He was very eager to make use of the intense neutron beams that existed at the Argonne reactors. He asked me to return to Argonne to work with him on these experiments. I had originally started working with Fermi at Columbia in 1941. In 1946 I had taken nine months off to obtain my Ph.D. In 1947 and the spring of 1948 Fermi and I drove out to Argonne National Laboratory two or three days a week. I would pick up the laboratory car at the motor pool and drive to his house, and then he would insist on being the driver or chauffeur. He was his usual very energetic self when driving. One day as we were driving south on Cicero Avenue, a slow-moving train was blocking our path. When he saw the caboose coming he gunned the engine; as the caboose passed, we went zooming across. There was a second set of tracks and a train on those tracks just missed us by a few feet. Fermi pulled over to the side so we could recover from our fright. He said to me, "This is why it is very important for you to be with me; my time may be up, but yours isn't." I assumed he was being jocular. Anyway, those of you who received your Ph.D.'s after 1947 owe the existence of your thesis adviser to my being with Fermi.

What were the experiments that we were doing at Argonne? I will come to the very major contribution to material science and other fields made by Fermi after I describe a couple of other experiments. One of them was that we wanted to build an atomic beam apparatus to measure the spins and magnetic moments of radioactive nuclei that might be produced in the high-flux reactors. It took longer than Fermi had anticipated; the only measurement that we made in 1948 was performed at the request of Maria Mayer, who was developing the shell theory of nuclei. She did not like the literature value for Na^{23} of spin 3/2; she wanted it to be 5/2. It was one of only two exceptions that she encountered to her very successful theory, which made use of Fermi's suggestion of L•S coupling.

Another experiment, this one with Leona Marshall, was looking for a neutron-electron interaction. The idea underlying the experiment is that one can picture the neutron as being a negative meson cloud outside a proton core, approximately 20% of the time. When the electron is inside the meson cloud, it should be attracted to the proton. The idea of doing this experiment is really due to Leona Marshall. The experiment consisted of sending a beam of neutrons through the many electrons in xenon gas. We were looking for a forward-backward asymmetry in the neutron scattering due to the interference between the scattering by electrons and the nuclear scattering. The apparatus was basically simple; it consisted of boron trifluoride neutron detectors measuring the neutrons that were scattered at 45 degrees and 135 degrees to the beam. The beam originated in the thermal column with apertures and a collimator made of cadmium and the final aperture at the face of the thermal column, to give a fairly well colli-mated beam of neutrons. We started to make measurements in Fermi's usual style of checking everything; we ran background measurements when the xenon was frozen out into a liquid nitrogen trap. We measured the fast neutron background in the beam by putting cadmium screens in the beam. We reversed the direction of the counters in the beam. The measurements indicated that something was wrong. How did Fermi proceed when he knew something was wrong?

We stopped making measurements and started discussing various tests we could try that would help us understand what was wrong. We tried the easy ones first and then the ones that might be the most probable cause of the experimental troubles. For example, a quick one would be to inter-change the electrical wiring on the detectors. We tried about five different things and several of them showed the trouble. The next time we came back, the measurements ran very smoothly and were completed in four hours.

As mentioned previously, in the neutron-electron experiment, we ran background for stray neutrons by emptying the xenon from the beam. In an earlier experiment I had made the astounding discovery that when we went to lunch so did some of the "stray neutrons." The background count-ing rates were lower at lunchtime. Fermi was skeptical that the neutrons went to lunch at the same time as we did. Fermi's alternative explanation was that when we were sitting next to our equipment, the stray neutrons were scattered from our bodies into our detectors. There went my astound-ing discovery.

Most of us became particle physicists and were not aware of Fermi's im-pact on the development of the techniques of neutron diffraction and scat-tering that were so important to the material science physicists as well as chemists and biologists. Fifteen years after Fermi's classic papers of 1947, there were forty research reactors around the world devoted to this tech-nology. He published a series of papers, mostly with Leona Marshall, but others were with Bill Sturm, Bob Sachs, and Herb Anderson. He measured scattering lengths (for twenty-two elements), coherent and incoherent scat-tering cross sections, the phase of the scattering, the spin dependence of the scattering, and isotopic differences. The techniques he employed included Bragg scattering from single crystals as either monochromators or spectrom-eters, a neutron velocity selector, total reflection from neutron mirrors, and molecular effects in neutron scattering by gases. Another very useful tool was the filtering of very slow neutrons by microcrystalline materials, such as BeO. Figure 6.5 shows a sharp cutoff of the short-wavelength neu-trons, which are scattered out of a beam while the long-wavelength neutrons pass on through. When the neutron wavelength is longer than twice the maximum lattice spacing they cannot coherently Bragg scatter. The microcrystalline material becomes transparent to the neutrons that have $\lambda > 2d$. A measurement of the neutron energy spectrum transmitted by the BeO was made by a neutron velocity selector. The neutron beam was chopped in time by using a collimator consisting of a sandwich of cadmium absorbers and an aluminum neutron-transmitting material. When the collimator was rotated at high velocity, neutron velocities from 500 to 5000 meters/second could be selected by the flight time. The velocity selector is described by E. Fermi, J. Marshall, and L. Marshall, *Physical Review* 72 (1947). Figure 6.6 shows Fermi checking the electronics of a velocity-selecting apparatus.

The interaction of long-wavelength neutrons with material can be de-

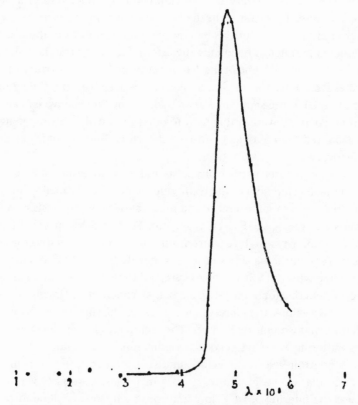

Figure 6.5 Neutron spectrum filtered through BeO. (Figure from E. Fermi and L. Marshall, *Physical Review* 71 [1947]: 666–667. Copyright 1947 by the American Physical Society.)

Figure 6.6 Fermi at the electronics of the neutron velocity selector in 1947. (Photo courtesy Argonne National Laboratory.)

scribed by an index of refraction. The nuclear interaction with the individual atoms averages to an interaction with the material as a whole. Fermi found many of the indexes of refraction to be less than one. Therefore, neutrons going from air onto a highly polished mirror will be totally reflected at small glancing angles. One morning, someone was looking for Fermi in Eckart Hall. We were told he was over in the machine shop at Ryerson Physical Laboratory. Fermi was sitting at the bench of a wonderful tool and die maker named Deconstanza (called Deke). He was busy showing Deke how to calculate the grit of the polish one needed to get the mirror surface satisfactory for the experiments. He loved showing people how to be quantitative. Deke asked him how he was going to check whether the mirror was satisfactory; Fermi said, "I'll hold the mirror up and if I can see my eyelashes in it, it will be OK." I have not figured out whether this was or was not a quantitative statement. The neutron mirror experiments consisted of a BeO filtered beam of neutrons with 1-millimeter slits at the end of the filter, and another 1-millimeter slit at the face of the thermal column. The mirror was on a stand, which could be adjusted vertically and tilted. A final slit in front of the neutron detector was several meters from the mirror. We tended to have our specialties: Leona usually made the detectors, Fermi took care of the mirror system, and I usually set up the neutron beams. Fermi loved these experiments; anyone involved had a beautiful demonstration of the wave and particle duality of quantum mechanics. The neutrons were mathematically waves when they were reflected from the mirrors. However, they were being detected as particles in the counters when they were captured by a boron nucleus, which then emitted an alpha particle. Figures 6.7 and 6.8 show the experiment. The mirror was raised until it was partly in the beam and then the mirror was tilted. The counters were moved vertically to find the totally reflected beam.

Owen Chamberlain was also a student of Fermi's who did his Ph.D. experiment at Argonne National Laboratory. His thesis was to measure the neutron diffraction by several liquid metals and to interpret the results to obtain the distribution of atoms in the liquids.

Beginning in 1948, Fermi and Bob Sachs arranged to hold a theoretical seminar at Argonne on Fridays. Those attending included Maria Mayer, Mort Hammermesh, David Inglis, and myself, all from ANL, Sid Dancoff and Maurice Goldhaber from Illinois, A. Seigert from Northwestern, (sometimes) Edward Teller from Chicago. We all took turns giving seminars on topics that Fermi suggested.

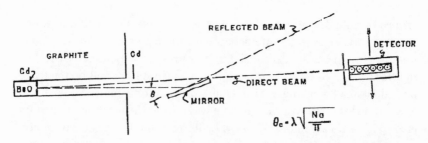

Figure 6.7 Arrangement for the study of mirror reflection of neutrons at the Argonne pile, 1947. (Figure courtesy private papers of Al Wattenberg.)

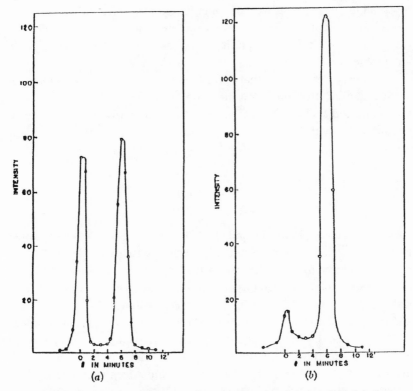

Figure 6.8 The direct and reflected neutron beams: *a*, the mirror intercepts half of the beam; *b*, the mirror intercepts the full beam. (Figure courtesy private papers of Al Wattenberg.)

During the period 1946–48, the students would very often eat lunch at a long table in the Hutchinson Commons. Fermi ate with us frequently. The students varied from time to time. The conversations dwelt on many things, including what we were doing. Fermi often would see something in the room that would lead him to give us a problem, such as, "How thick can the dirt be on the window up there before it falls off?" To solve the problem, one needed to know the fundamental constants of nature and something about the forces involved. Another problem he gave us, which was more sophisticated mathematically, came from observing the flames in the fireplace, which were going up the chimney. The pressure must be lower above the fire. The questions were, "What is the pressure above the fire?" and "How does it vary with the height of the chimney?" That was a problem which required you to go from a second-order differential equation to a linear differential equation. We were not using pencil and paper;

it was all in our heads. If we said something that seemed wrong to Fermi, he would ask us to go to an extreme condition to see if we had given a ridiculous result. It was lots of fun! Fermi made us feel that we could solve any problem.

I was very fortunate in that Fermi enjoyed being with me, sometimes as a competitor. He would beat me at tennis. I almost always beat him at Ping-Pong and at chess, and he was a much better swimmer than I was. It was fun being with Fermi . . . except when he was my chauffeur.

❖ ❖ ❖

Courtenay Wright
FERMI IN ACTION

I arrived in Chicago in the late fall of 1949, one of Berkeley's newly minted Ph.D.'s. My sponsor, Emilio Segrè, from Enrico's 1920s and 1930s Roman cadre, introduced me to Enrico, and, armed with a National Research Council fellowship, off to Chicago I went. It was the start of a wonderful but tragically all-too-short experience for me. Berkeley, with its 184-inch cyclotron and the great pion experiments of Herb York, W. H. K. (Pief) Panofsky, Jack Steinberger, Ceasar Lattes, and many others was exiting, but being around Enrico was something else entirely. His working habits were exemplary: he arrived at 8 a.m., when many of us were sound asleep from overwork or something else. He left at 5 p.m. sharp, with the possible exception of a couple of times when our local cyclotron was coming on line. Looking at his performance during working hours I am reminded of a tank: a steady and irresistible vehicle moving moderately; but God help any physics problem, simple or abstruse, that had the temerity to get in the way. He was generous to a fault and seemed to have a special feeling for the young people around him. He would regularly go with us to a small table at the Quadrangle Club for lunch and engage in wide-ranging discussions on subjects often brought up by him. For a year or two he had a number of us meet in his office in Eckhart Hall once a week in the evening to "talk" about a wide range of physics problems. The "talk" was mostly a lecture by Enrico in his inimitable style. There were also weekly meetings in Gregor Wentzel's office to work over modern problems. It was there I heard firsthand accounts of his cosmic ray acceleration method driven by moving magnetic fields and his thermodynamic view of multiple particle production in energetic collisions. Incidentally, the density of cigar smoke in Gregor's office would now be considered at toxic dump levels.

There were a number of seminars—Joe Mayer's institute seminar, still meeting today, was remarkable for its variety of topics, and its informal style: speakers were volunteers or chosen on the spot by Joe pointing a finger at you and asking you to talk—it was not advisable to arrive with nothing but a vacant mind. I think it a shame that grad students were excluded. This was understandable early on: my first attendance was in a classroom in Kent Chemical Laboratory, where I squeezed into the back row, but when we moved over to the institute, grad students could certainly have been fitted in, no doubt in some back-row ghetto. The participants, as well as the topics, covered a broad range: Bill Libby, the father of carbon dating; Harold Urey, then studying the history of the early solar system as recorded in meteorites; Harrison Brown, an early geochemist; Sam Allison, creating the northern lights in his laboratory; Nathan Sugarman and Tony Turkevich, both superb nuclear chemists; Subramanyan Chandrasekhar, keeping us up to date on the goings-on in the cosmos; Maria Mayer, then developing her shell theory of nuclei; occasionally Leo Szilard; and many others. It was quite a crew, most brought here by Willy Zachariasen.

Enrico had a wonderful capacity for contributing to problems as they emerged. I could not confirm this, but I believe it was his suggestion that the cyclotron magnet coils, 20 feet in diameter, made of 2×2-inch copper, be cooled by water flowing within them rather than the then-universal practice of bathing them in circulating oil. It immediately became the world standard for large coils. When our 450-MeV cyclotron, "High Energy" in those days, started spewing out pi-mesons in large quantities, Enrico made beautiful use of C. Störmer's treatment of the motion of charged particles in the axially symmetric magnetic field of the earth. His variant described the focusing and energy dispersion of pions by the fringing field of the cyclotron. This was of critical importance for the scattering experiments then ramping up. It was at that time that he constructed his famous cart that held a beryllium target. It ran around the rim of the 170-inch-diameter magnetic pole pieces with a motor employing the cyclotron magnetic field, and rotatable coils attached to the cart and energized externally—all made by the master in his lab, using an old South Bend lathe.

Let me close with a vignette that is dear to me. You need to know that in informal seminars, if the speaker—it could be anyone—did not thoroughly understand his subject and convey this to the audience, he was in danger of Enrico interrupting and completing explanations in high style. My tale occurs at our small table in the Quadrangle Club, where, after finishing lunch, Enrico reported on a recent trip to Los Alamos, where he collaborated with the Hungarian mathematician and computer master Johnny von Neumann on some undisclosed problem. As Enrico reported, Johnny was calling on obscure changes of variables, unheard-of transforms, and, for all I remember, the Heimlich maneuver, in a tour de force solution of the problem, completed in jig time. Enrico turned to us with a quizzical look on his face and said, "You know, I felt like the fly who sits on the plow and says, 'We are plowing.'" Wreathed in smiles, we all fell off our chairs.

Chapter Seven

Reminiscences of Fermi's Students, 1945–1954

❖ ❖ ❖

Harold Agnew

A SNAPSHOT OF MY INTERACTIONS WITH FERMI

In January 1942 I went to the University of Chicago to join the Manhattan Project. I was immediately sent to Columbia University to work with Enrico Fermi. When I first met him, the only unusual thing that I noticed was that all of his pants pockets had zippers—all four of them.

At the time he was conducting experiments using a large pile of graphite. The structure was entirely encapsulated with a sheet metal cover and was evacuated using mechanical vacuum pumps. The pile had a radium-beryllium neutron source at its center, and we measured the slowing down of the neutrons using indium foils which were activated by the source's neutrons. We would insert the foils at different levels in the pile for a specific time, then remove them and run about 100 feet to the counting room, where there was a set of Geiger counters. We did this hour after hour for about ten hours each day. Fermi not only directed the work, but actually took on a shift the same as the rest of us, inserting the foils, running to the counting room with the activated foils, and then taking the data.

He was one of us. This always distinguished Fermi. He clearly was a genius, but acted with no pretentiousness. He was a very unassuming person. He had a wonderful sense of humor. The array of counters in their lead shields all had names, taken from the *Winnie the Pooh* books. They were named "Pooh," "Piglet," "Heffelump," and so on.

For nonnuclear safety reasons he decided to move the experiments to Chicago, and we started to build CP-1, the first man-made chain reaction. One day a several-ton load of graphite blocks was delivered around 4 p.m. We had to unload the truck, so along with the rest of us, Fermi took off his coat and pitched in and helped unload the truck. This was Fermi. He not only supplied the brains at Chicago, but when needed also supplied the brawn.

Chicago is cold in the winter, and people went ice-skating there near the university. Fermi had never ice-skated and decided he would. We all went to the rink, got Fermi a pair of skates, and after a few falls, Fermi caught on, and before the end of our first session was skating as well as anyone else.

He was an excellent athlete and loved to compete. He liked to play tennis especially. Later on, when I returned to Chicago as a graduate student, we used to play tennis during the lunch hour. This required checking out a net and setting it up on the court. The professor and student took turns with this task. He was a very regular person, not at all impressed with his position.

The only sport at which he was a failure was in fly-fishing for trout. Emilio Segrè, who was a very good fly fisherman, never let Fermi forget that at this sport he was no good.

In 1946, after the war, housing was very scarce in Chicago. I was unable to find a place for my family to live. Fermi, who had a fairly large house, suggested that my wife, small daughter, and I come and live with them. His wife, Laura, wanted to visit her sisters in Italy, and when she was gone, my wife, Beverly, could run the house and do the cooking and so on for Fermi and his children, Giulio and Nella. We did this for almost three months, until I found a place for us to live. Being part of the family for three months was a wonderful experience. Fermi preferred nonspicy food and always diluted the red wine we had for dinner half with water. We stayed on for a month or so after his wife Laura returned.

One evening she told Fermi that she had gone to the local appliance store and put her name on a waiting list for a General Electric dishwasher. (After the war appliances were scarce and one had to sign up on waiting lists for appliances, cars, and such.) I was astounded. Fermi had been the major consultant for General Electric, which was building reactors at Hanford for the production of plutonium. I said, "Enrico, you know the president of General Electric. Just tell him you want a dishwasher and he will send you one tomorrow." Fermi thought for a second and said, "No, that wouldn't be fair for others, we will wait our turn in line." This was classic Fermi.

Fermi liked to swim. Sometimes after work his team, of which I was a member, would go to Lake Michigan. On one day he decided we would swim across a little bay. I had been a varsity swimmer in high school, so thought I was pretty good. But after about fifteen minutes in the choppy, cold water of Lake Michigan, I was falling behind. Fermi, who swam with what I would call a "dog paddle" style, swam back to me and asked if I was OK. I said I thought so, but clearly my Australian crawl swimming style wasn't best for choppy Lake Michigan. I barely made it to the other side of the bay, and with difficulty climbed up the sea wall and sat down. Fermi said, "Meet you back where we started," and plunged back in and swam back to our starting point. I had difficulty just walking back.

Fermi was known by his colleagues as "the pope." This made it all very clear that he was the supreme authority on all matters. He held this position in all of our minds as an accepted fact—no big deal, just an accepted realization that he really knew more than the rest of us, or anyone else involved in our scientific work.

Fermi especially liked young people. He, in his position, entertained a lot, but preferred to have young people. The top floor of his Chicago house had a large room in which he would invite students to come and square dance. I usually did the calling, and a good time was had by all. He and Laura had these parties about once a month. When he had dinner parties for his peers, he always said, "We need to dilute 'So-and-so' and 'So-and-so' with some young people; the 'So-and-sos' are too stuffy."

Chicago had an open enrollment system for graduate studies but required a three-day written examination to decide one's future. Choices were "flunk and out," "pass with a master's degree and out," or "pass with option for going on for a doctorate," if you could find a faculty sponsor. I was terrified about taking the exam because I felt my peers were much smarter than I. (Subsequently, four of my classmates have received a Nobel Prize in physics, and they were not all the really smart ones.) The tests were given so that those scoring the written results had no idea as to whose papers they were grading. I kept putting off taking the test, but Laura Fermi kept urging me to take it. I went to Fermi and asked what he suggested I read. He said he had no idea because he didn't read much. I asked how he always knew what was going on. He said people came and told him and explained things to him. Then he said, and it amazed me, that there were people who said they immediately understood things but he wasn't one of those. He said it took him a long time to understand what people were explaining to him, but many times he realized that they really didn't understand what they were describing to him, but he did. He also volunteered that one who was very quick to say he understood even before the person finished was Robert Oppenheimer, but a lot of the time, Oppenheimer really didn't understand the technical information the way Fermi understood it. He told me that if you really understood (Fermi's way of understanding) about ten things in physics you could know almost everything.

I had been getting a week's lecture on Brillouin zones, which I never understood, and asked him about it. He went to a small blackboard, and in less than five minutes developed the whole theory, and at the time I thought I understood it. But as it was with most of Fermi's lectures, they were so clear and so simple that you really thought you understood all, but when one tried to repeat it afterward on one's own, one became lost. Very much like eating Chinese food: "end up very full and satisfied but shortly very empty and hungry."

Of all his colleagues of his vintage, Fermi's favorite for his intellectual ability was Edward Teller. He told me this, and years later Laura Fermi and his daughter confirmed this when I raised the question. Among his young people, I believe Fermi thought Dick Garwin was the brightest, and I also believe this even to this day.

This is just a snapshot of my interaction with Fermi.

❖ ❖ ❖

Owen Chamberlain

A BRIEF REMINISCENCE OF ENRICO FERMI

My First Meeting with Enrico Fermi—Los Alamos

My first meeting with Enrico Fermi was remarkable for its low key. I had been looking for Emilio Segrè; I found Segrè in a small shop equipped with an electric drill press. As I turned to leave, Segrè said, "Oh, Chamberlain, I want you to meet Fermi." My mouth dropped open.

These reminiscences were written when Owen Chamberlain was quite ill, and they required a great effort on his part. See also the letter of Owen Chamberlain to Fermi in chapter 4.—Ed.

I knew what a Nobel Prize winner looked like, and Ernest Lawrence fit the bill exactly: big, with a voice that echoed down the hall. Here was this little man, sitting motionless in the corner of the very small room.

On the Train

I had been in Los Alamos less than three months when Oppie (Robert Oppenheimer) asked me to go on a trip to Ohio to see if reasonable progress had been made in forming a radioactive source. (In those days you had to change trains in Chicago.)

Imagine my delight when I found in the compartment next to mine—none other than Enrico Fermi!

I soon had him answering questions and doing problems.

Fermi in the Hospital

When it became known in Berkeley that Fermi had an incurable stomach cancer, Segrè turned his classes over to me and, with Subrahmanyan Chandrasekhar, went to Fermi's bedside in Chicago.

There was bound to be a moment of tension—until Fermi said: "Chandra, do you think I could be an elephant the next time around?"

❖ ❖ ❖

G. F. Chew
PERSONAL RECOLLECTIONS FROM 1944–1948

I first encountered Enrico Fermi at Los Alamos in 1944. In January of that year I had been shipped to the supersecret laboratory at age nineteen with a bare-bones wartime B.S. degree in physics, to be an assistant to Edward Teller. Although I had probably heard of Fermi from George Gamow, my influential teacher at George Washington University who had responded to a request from Teller by shipping me off, I had no appreciation of Fermi's accomplishments until older Los Alamos colleagues began to recount them to me. My command of physics, including its history, was abysmal.

Inside the close-knit "white-badge" Los Alamos community allowed to share the secrets of the Manhattan Project, I got to know Enrico and his wife Laura on a first-name basis and became aware of a Fermi personality that, while good-humoredly gentle and considerate of others, was intensely competitive. Trivial pursuits, such as hiking and party games, were addressed with dedication. Fermi never enjoyed being second best. He was systematic about anything he did. Enrico decided systematically to Americanize himself by adopting our peculiar customs. Noticing our penchant for nicknames, he did some research and one day announced that henceforth we should call him "Hank." (Nobody ever did.)

At war's end, Los Alamos organized a semester of graduate courses to occupy scientific staff until they returned to their various university homes. Fermi elected to give a course in nuclear physics and asked me to be his TA. I was dumbfounded, never having attended a graduate course of any kind and being keenly sensitive to (and embarrassed by) my ignorance of quantum mechanics. It was nevertheless an offer that could not be refused, and Fermi's legendary technique of teaching allowed me to survive. I was introduced to quantum mechanics through down-to-earth nuclear physics applications that ignored philosophical issues and focused on results.

Having to grade the homework problems and exams in Fermi's Los Alamos course may have been my decisive educational experience. Fermi's awesome talents as a teacher need no documentation from me. I shall only remark that he never lost patience with a student who failed to grasp Fermi's first presentation of an idea. Nor did he mind if a second or third presentation continued to elude. It appeared as though Fermi welcomed the opportunity to repeat the pleasure he got from a physics exposition.

It was understood at the University of Chicago, where I followed Teller and Fermi in 1946, that Edward Teller was to be my research supervisor. But in mid-1947, after completion of all my exams, Teller told me that Fermi, while waiting for construction of the Chicago cyclotron to be completed, had decided to take two theoretical students to help himself catch up on recent theoretical developments in "high-energy physics." Teller asked if I were interested. I dashed down the hall to Fermi's office and signed up. A few moments later I encountered Murph Goldberger and told him of my luck. He asked how many theoretical students Fermi wanted. I said, "Two," and Murph was off in a flash.

The pair of us worked intensively with Fermi during 1947–48, meeting every day for lunch at the commons and spending much of the afternoon at Fermi's office in drawn-out discussions. In retrospect, the amount of attention Murph and I got from Fermi is unbelievable. Many questions were examined and many calculations performed. One fine day in mid-1948 Fermi said it was time for us to move on and to become independent. There had been no talk of publishing any outcome of our work, but Fermi now reluctantly recognized the need of conforming to university regulations. In order to throw us out into the world, there needed to be two separate published theses. Fermi said we should somehow carve out two portions of the aggregate work done together and write papers that could be published.

The segregation of material was totally artificial but somehow accomplished. I wrote and published in *Physical Review* a paper under my name alone, even though the work was a collaboration with Murph and Enrico, based on Fermi's ideas. Fermi had no concern for "credit," being interested only in results. Throughout my subsequent academic career, Fermi's attitude prevented me from ever considering a doctoral thesis as evidence of anything beyond ability to follow arbitrary rules.

I have alluded to Fermi's pragmatic attitude toward quantum mechanics. He possessed an instinctive grasp for the relation between theory and the physics meaning of "truth"—a meaning based neither on mathematics nor logic, but on the Galilean idea of reproducible experiment. Because all measurements have errors, such a meaning for truth is slippery and changes as experimental techniques are enlarged and refined, but the meaning is the foundation of natural science. Theoretical physics has repeatedly generated beautiful mathematical structures to correlate certain categories of experimental measurements, only to find these esthetic marvels unable to cope with all measurements. Fermi, who was at once theorist and experimenter, possessed an extraordinary sense for the relation between the physical world and its mathematical portrayal by theorists—an attitude whose meaning for Fermi's 1948 theory students Murph came to summarize by the statement, "Since it is true it can be proved." Others see in Murph's summary a joke. I find it a profound appreciation of Fermi's special gift.

I had further contact with Fermi after I left Chicago, but there was nothing to match the interaction experienced between 1944 and 1948. I happened to stumble into an interval of Fermi's late career, between slow neutrons and pion-nucleon scattering, when he was not absorbed by experiment and was getting his kicks from teaching. What extraordinary luck!

❖ ❖ ❖

George W. Farwell
REMEMBRANCES OF FERMI

After all that has already been written about Enrico Fermi, perhaps the best I can do is relate my own perceptions of him and my experiences with him at Chicago and earlier, at Los Alamos, New Mexico, during the latter years of World War II.

I nearly chose as my title "Enrico Fermi: A Generous and Kindly Genius," but it seemed a little too poetic. That he was truly a genius, we all know. But to a young graduate student he seemed also to be incredibly generous in sharing his time, his talent, and his vast knowledge of physics with a group of young people who had little to give him in return, unless it was the hope that some or all of them would actually accomplish something significant in science. It is sad that he didn't live long enough to witness the Nobel Prizes and other honors that later came to a number of his students.

I say "kindly" for several reasons. Over a period of several years both at Los Alamos and at Chicago, I never once heard him respond caustically to a question or comment, or embarrass someone with an intellect clearly inferior to his own. (Of course, that included almost everybody.) I cannot say the same for some of the other prominent physicists of that time. And he and his lovely wife Laura often entertained his young people and their spouses, graciously and generously, in their home near the university. Also, to me personally, he was a very kindly guide and adviser.

Until June 16, 1943, Fermi had been, to me, but a legendary figure in the exciting physics of the time. On that day, en route to Los Alamos from Berkeley, I rode with Emilio Segrè, my immediate boss at Berkeley and the leader of our new group at Los Alamos, on a DC-3 very bumpily bound from Oakland to Albuquerque. For several hours (until he became too airsick to continue), Segrè told me about his early experiences in Rome with Fermi; about Fermi's pile here at Chicago, which had gone critical only a few months before; about the new element plutonium and its nuclear properties; and about the plan, at Los Alamos, to make an atomic bomb as quickly as possible.

During 1943 and 1944, at Los Alamos, Segrè was always in close touch with Fermi at Chicago, and he kept telling us that Fermi would soon come to Los Alamos (which he did, in the summer of 1944). Segrè, who must have been all of five feet seven inches tall, also said of Fermi—and if he said it once he said it twenty times—"You know, Fermi is just a *little* guy!" Well, here is a picture of Fermi and Segrè at the edge of the Valle Grande, a vast volcanic caldera in the Jemez Mountains near Los Alamos (fig. 7.1).

Segrè's group at Los Alamos, and especially Clyde Wiegand, Owen Chamberlain, and I, were given the task of measuring spontaneous fission in uranium, plutonium, and other heavy elements in order to ascertain whether neutron emission from spontaneous fission could become a factor in the design and, ultimately, in the assembly of an atomic bomb. (Too many neutrons from spontaneous fission would start a premature chain reaction during assembly and lead to a fizzle rather than an explosion.) Because of the near-zero counting rates expected, these experiments were carried out

Figure 7.1 Fermi and Emilio Segrè at Los Alamos, 1944. It's hard to say who's little and who isn't! (Photo courtesy Lawrence Berkeley Laboratory.)

in an isolated location—an old Forest Service log cabin in Pajarito Canyon, a few miles south of Los Alamos.

Early in 1944 we began to find increasingly high spontaneous fission rates in pile-produced plutonium, with a definite correlation between the observed rate and the total irradiation that had been received by the uranium from which the plutonium was extracted. Fermi made a crucial suggestion: take a little pile-produced plutonium (a few milligrams sufficed), expose it further to radiation in the pile, and observe the result. The result was clear and, for the plutonium bomb project, stunning: the greatly increased spontaneous fission rates observed in reirradiated plutonium were very much too high for a "gun" assembly (the simplest approach for a bomb) to be feasible for plutonium. The increased rates were almost certainly due to Pu-240, a new isotope created in the pile by neutron absorption in Pu-239.

Although some loose ends about the spontaneous fission in plutonium remained to be nailed down during the ensuing year, on July 25, 1944, Robert Oppenheimer told the assembled scientists at Los Alamos that the plutonium gun would be abandoned and an all-out effort begun to make a much faster acting "implosion" device. As we all know, this colossal effort succeeded, and the first implosion device was detonated, not quite a year later, on July 16, 1945, in the New Mexico desert.

The discovery of Pu-240 in 1944 had immediate consequences here at Chicago, too: the massive plutonium ultrapurification program at the Metallurgical Laboratory under Arthur Compton was seen to be unnecessary, since the neutrons from alpha-induced reactions on impurities would no longer be significant compared to those from spontaneous fission.

Upon being reminded by Segrè that his crucial suggestion had borne fruit, Fermi couldn't remember having made it! Perhaps this simply demonstrates that he was only human, after all.

In late 1945 and 1946 there was a general exodus from Los Alamos; some faculty and graduate students returned to their former institutions, but many others elected to make a change. (I believe that this was part of the greatest shuffle ever experienced by the scientific community.) Segrè went back to Berkeley but passed me along to Fermi, who was headed for Chicago and the new Institute for Nuclear Studies. Fermi very kindly agreed to be my sponsor here and to accept a major part of the plutonium work already accomplished as my Ph.D. thesis project. Also—on a postwar fishing trip in the Rio Grande box canyon below Taos—Sam Allison offered me a fellowship in the new institute. So it all came together, and I came to Chicago in the spring of 1946.

All was not peaches and cream with the thesis work, however. Fermi did not like loose ends, and he made me learn enough pile theory to demonstrate, using the best available data on neutron fluxes and distributions in the pile and on the relevant cross sections, that we had observed the expected amount of Pu-240 production. (Fortunately, it all checked out.) This is perhaps why my thesis remained classified Secret until about 1959, so that I never had the pleasure of defending it publicly; Fermi, Allison, and two other professors had at me in private, though, and somehow let me through.

But the experience at Chicago during 1946–48 included also an unexpected and priceless component: a small group of us met about one evening a week with Fermi in his office (occasionally in his home) in an unplanned, unstructured seminar in the course of which Fermi simply took off in one or another direction and we all followed, participating in whatever way we were able. The range of topics was broad, touching at one time or another almost every branch of physics, and Fermi's lucid and intuitive

Figure 7.2 Fermi in his office in 1945. (Photo courtesy Los Alamos National Laboratory.)

approach was always a pleasure to behold and to experience. I have always been most grateful for this wonderful, continuing exposure to one of the great minds of our time. As best I can recall, in 1946 and 1947 our little group included Harold Agnew, Harold Argo, Owen Chamberlain, Geoffrey Chew, Joan Hinton, Leona Marshall, and C. N. "Frank" Yang. (If I have inadvertently missed someone, I apologize.)

Finally, here is a picture of Fermi in 1945 in his modest office at Los Alamos (fig. 7.2). (All offices in wartime Los Alamos were modest!)

To sum it all up: I often ask myself, thinking back more than half a century to my experience with Fermi, "Hey, how lucky can you get?"

The answer: very lucky indeed.

❖ ❖ ❖

Uri Haber-Schaim
FERMI IN VARENNA, SUMMER 1954

Fermi's hundredth anniversary is a double one for me; it is also fifty years since I received my Ph.D. under Fermi's guidance after receiving my M.Sc. at the Hebrew University under Fermi's former student, Giulio Racah. I was privileged to spend part of Fermi's last summer in Varenna with him. My comments about that summer are intended as captions to a few photographs, which I will share with you.

Lectures by the senior staff were in the morning; seminars by participants in the afternoon. All the lectures and seminars were published in the *Supplemento del Nuovo Cimento* number 2 (1954). Fermi's lectures, edited by B. Feld from notes taken by the participants, constitute the last paper in Fermi's collected works.

Before the opening ceremony, the garden in front of the villa was full

Figure 7.3 A photograph of Laura and Enrico Fermi at the villa which appeared in the Italian magazine *Oggi*. Uri Haber-Schaim is shown with has his back to the camera. (Photo courtesy RCS Periodici, Milan, Italy.)

Figure 7.4 Relaxing on the steps of the Villa Monastero. *Left to right:* Werner Heisenberg, Fermi, M. Bruin from Holland, Louis Leprince-Ringuet, and Bruno Rossi. (Photo courtesy Uri Haber-Schaim.)

of "paparazzi" as in *La Dolce Vita*, waiting to take pictures of Fermi (fig. 7.3). However, this being only Fermi's second visit to Italy after the war, they did not know what he looked like. After Enrico and Laura appeared, and I walked over to greet them (not having seen them for three years), the cameras started clicking. I moved a little to the side and we pretended not to take notice of all the commotion. A few days later, one of the participants told me that my picture appeared in *Oggi*, a pretty bad Italian weekly. I bought a few copies.

During breaks and before dinner, relaxed shoptalk and plain socializing took place in the garden between the Villa Monastero and the lake, and in particular on the steps leading to the water. Figure 7.4 shows such a gathering. One of the participants was an attractive lady who used to go swimming in the afternoons wearing a minimal bikini, her hair loose, and flippers on her feet. She definitely was part of the scenery. One day Enrico asked her to pose for him. I grabbed a camera from a friend and took the picture shown in figure 7.5

Most of the participants stayed at the same hotel. After dinner, much of the socializing was done in the Calcetto (table football) room (fig. 7.6). The senior lecturers often came over, too. For those of you not familiar with the game, suffice it to say that there are between two and four players on each team manipulating four rods on which the eleven "soccer players" are

Figure 7.5 Photographing the photographer. (Photo courtesy Uri Haber-Schaim.)

mounted. The game had become very popular among particle- and cosmic-ray physicists during the Bagnères-de-Bigorre conference in 1953. Some players had great skill in passing the ball and shooting it across the table straight into the opposing team's goal. Enrico barely ever played the game, but he agreed to play one evening. His strategy, especially when he was on defense, was to apply statistics. He whirled his rod with great zest to increase the probability that one of his men would hit the ball. It did not work very well. Figure 7.6 shows Fermi and colleagues at the table.

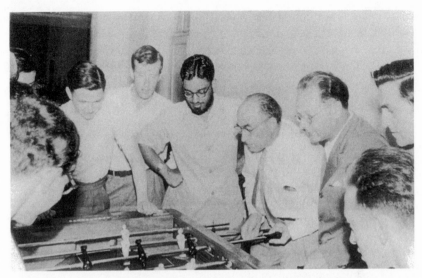

Figure 7.6 Fermi (*fourth from right*) at the football table. *Left to right around the table:* Marcello Conversi, Denis Keefe, George Clark, M. K. Menon, Eduardo Amaldi, G. Puppi, and an unidentified colleague. (Photo courtesy Uri Haber-Schaim.)

Shortly after the course I sent Enrico a bunch of photographs from the course, including the one where he had photographed a lady in a bikini (fig. 7.6). A few weeks later I heard over the Swiss radio that Fermi had died. I had no idea whether he had seen the photos. Having heard the sad news, I wrote to Mrs. Fermi, and in due time I got the following note written on her calling card:

> Your pictures came in time for Enrico to enjoy them. Thank you for them and for the kind lines you wrote me now. Sincerely, Laura Fermi.

❖ ❖ ❖

T. D. Lee
REMINISCENCE OF CHICAGO DAYS

In 1946, Fermi joined the faculty of the University of Chicago. The same year, I received a Chinese government fellowship which enabled me to come to the United States to further my study in physics.

At that time, I had only two years of undergraduate training in China. Although I did not know much English, I was already familiar with classical physics and knew some quantum mechanics. I felt well prepared for graduate study. But at that time to enter graduate school without a college degree was almost impossible, except at the University of Chicago, which was willing to take people without a formal degree provided they had read the great books of Western civilization selected by President Hutchins. However, I had zero knowledge of any of these great books. Luckily for me, I had the help of the Chicago physics department. Apparently, Fermi, William Zachariasen, John Simpson, and other professors convinced the admissions officer that I was quite knowledgeable in the oriental equivalent of such classics (Confucius, Mencius, Laotse, etc.), which she accepted. (I am grateful that years later I was told by Dr. Simpson of Fermi's role in getting me admitted to the University of Chicago. At that time, I only knew that the physics department had helped.)

Right after the war, the Chicago physics department was the best in the world. Besides Fermi, there were S. Chandrasekhar, J. Mayer, M. Mayer, R. Mulliken, J. Simpson, E. Teller, H. Urey, and W. Zachariasen (later G. Wentzel also joined the staff) on the faculty. I was indeed happy to be admitted as a graduate student at Chicago.

The first thing I did after my arrival was to read the university catalog. As I recall, it said the Department of Physics was only interested in exceptional students. It did not encourage students to take courses; however, for those who needed guidance, courses were also provided. I thought to myself that that was really the proper style of a great university. How unlike Southwest Associated University in Yunnan, where students absolutely had to take courses. Nevertheless, since Fermi was not scheduled to give any courses that quarter, I did register for quantum mechanics with Teller, electromagnetic theory with Zachariasen, and, later, statistical mechanics with both Mayers. By attending those classes I felt I was betraying the secret that I was not an exceptional student. However, that feeling was soon dissipated by my observing that there were many other students in these classes.

A few weeks later, I received a note from Fermi asking me to attend his special evening class (which was by invitation only). It was there that I had my first glimpse of Fermi in action. The subject matter ranged over all topics in physics. Sometimes he would randomly pull a card out of his file, on which usually a subject title was written with a key formula. It was wonderful to see how in one session Fermi could start from scratch, give the incisive estimate, and arrive at the relevant formula and the physics that could be derived from it. The freedom with which he moved from field to field was an inspiration to watch.

At one time, Fermi happened to pull out his index cards on group theory, which contained only titles, all listed alphabetically. He then started to lecture on *Abelian group* first, *affine correspondence* second, *central of a group* third, and then *character of a group*, and so on. Some of us were a bit confused by this unorthodox approach. Fermi said, "Group theory is merely a compilation of definitions, and alphabetical order is as good as any." In spite of his assurance, students like myself, who were not able to apply the permutation group fast enough to change the order, had some difficulty.

Another thing which comes frequently to my mind about student life in those times is the square-dancing parties held frequently at the Fermis' home. They were my first introduction to occidental culture. Enrico's dancing, Laura's punch, and Harold Agnew's energetic calling of "do-si-do" all made indelible marks on my memory.

Soon after, I became Fermi's Ph.D. student. Besides me, Dick Garwin was also Fermi's student, but he was an experimental physicist (and still is an extremely brilliant one). At that time, most of the Chicago Ph.D. students in theory were supervised by E. Teller; those included M. Rosenbluth, L. Wolfenstein, and C. N. Yang.

The relation between Fermi and his students was quite personal. I would see him regularly, about once a week. Usually we had lunch together in the commons, often with his other students. After that, Fermi and I would spend the whole afternoon talking. At that time, Fermi was interested in the origin of cosmic radiation and nuclear synthesis. He directed me first toward nuclear physics and then into astrophysics. Quite often he would mention a topic and ask me if I could think and read about it and then "give him a lecture" the next week. Of course, I obliged and usually felt very good afterward. Only much later did I realize that this was an excellent way of guiding the student to be independent.

Fermi fostered a spirit of self-reliance and intellectual independence in his students. One had to verify or derive all the formulas that one used. At one point I was discussing with him the internal structure of the sun; the coupled differential equations of radiative transfer were quite complicated. Since that was not my research topic, I did not want to devote too much time to tedious checking. Instead, I simply quoted the results of well-known references. However, Fermi thought one should never accept other people's calculations without some independent confirmation. He then had the ingenious idea of making a specialized slide rule designed to deal with these radiative transfer equations

$$dL/dr \propto T^{18} \text{ and } dT/dr \propto L/T^{6.5}$$

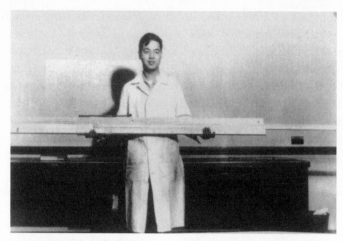

Figure 7.7. T. D. Lee holding a dedicated slide rule for the calculation of the internal temperature distribution of a main-sequence star, c. 1948. (Photo courtesy T. D. Lee.)

(where L is the luminosity and T the temperature). Over a week's time, he helped me to produce the magnificent six-foot-seven-inch slide rule shown in figure 7.7, with *18 log x* on one side and *6.5 log x* on the other. With that, even integration became fun and I was able to complete the checking quickly and then move on to a different topic for discussion. This unique experience made a deep impression on me. Even now, sometimes when I encounter difficulties, I try to imagine how Fermi might react under similar circumstances.

In 1948, Jack Steinberger, another student in Fermi's lab, was doing experimental work on the e-spectrum in the decay of $\mu \rightarrow e+$ . . . We talked a lot about his measurements, which indicated that μ-decay, like β-decay, involves four fermions. I became quite interested in that, and so did Rosenbluth and Yang. Are there other interactions besides β-decay that can be described by Fermi's theory? The three of us decided to make a systematic investigation.

We found that if μ-decay and μ-capture were described by a four-fermion interaction similar to β-decay, all their coupling constants appeared to be of the same magnitude. This began the universal Fermi interaction. We then went on to speculate that, in analogy with electromagnetic forces, the basic weak interaction could be carried by a universally coupled intermediate heavy boson, which I later called $W^\pm$ for "weak." Naturally I told Fermi of our discoveries, and he was extremely encouraging.

One serious difficulty that faced us was how to generate such a universally coupled intermediate boson from symmetry considerations. In order to have short-range interactions and to escape detection, the boson must be massive and unstable, with a very short lifetime. However, other universally coupled quanta, like photons and gravitons, are all massless and stable. In addition, because of parity conservation, it was difficult at that time to understand why there were both Fermi and Gamow-Teller interactions in β-decay. I made no progress in this direction, and procrastinated in writing it up. At the end of December 1948, Fermi called me into his office and said he had just received two reprints by J. Tiomno and J. A. Wheeler, which also discussed the universal weak interaction. He insisted that I should immediately publish whatever I had; furthermore, he would send a copy to Wheeler with a letter saying it was independently done some months earlier, which he did. I was quite touched by his thoughtfulness.

In one of the weekly afternoon sessions with Fermi, I mentioned the work by R. E. Marshak on white dwarf stars, which was based on a suggestion by H. Bethe. In the original Chandrasekhar limit, the critical mass of a white dwarf star was set to be $5.75 \, \eta^2$ times the solar mass M, where η is the ratio of electrons to nucleons in the star ($\eta = 1/2$ for a helium star and 1 for a hydrogen star). Marshak found that because of the high electronic conductivity in a white dwarf, its interior temperature can be quite low; he was able to produce an acceptable solution for a white dwarf consisting of pure hydrogen. That would make the critical mass 5.75 times the solar mass M. In the course of our discussion on the Marshak paper, Fermi asked me, with his usual insight, whether anyone had studied the question of stability. Soon I found that no one had. Then, I was able to show that the Marshak solution was unstable. Consequently $\eta = 1/2$, and the critical Chandrasekhar mass limit for the white dwarf is $1.44 \, M$, not $5.75 \, M$. This became my Ph.D. thesis, which I completed at the end of 1949.

❖ ❖ ❖

Jay Orear
MY FIRST MEETINGS WITH FERMI

My first course with Fermi was Quantum Mechanics, taken in the fall quarter of 1947. I was just one face out of almost one hundred. But I really met him in a more unconventional way. That same quarter I also had registered for a physical education course called Social Dancing. Early in the course one of the coeds in the class invited me to a dance party at a girl-friend's house. As we were walking to the house that night she happened to mention the name of her girlfriend as Nella Fermi, an art major. I asked whether her friend was the daughter of *the* Fermi. Being an art major, my date had never heard of Enrico Fermi and did not know. But once I entered the door, I was greeted by the warm, smiling face of my quantum mechanics instructor. (This was my first course with Fermi.) I was surprised that Fermi recognized my face and he asked me what I thought of his quantum mechanics course.

The party was a square dance with Harold Agnew as the caller. Many were Nella's friends (mostly female) and Enrico's coworkers (mostly male). I was an indirect guest of Nella and not Enrico. I was invited as a friend of a friend of Fermi's daughter. These Fermi square dances were held once a month. Fortunately, I was better than the average square dancer. From then on I was on the guest list of the Fermi family. The guest list was worked out by Nella, Laura, and Enrico. Harold Agnew did the calling and supplied the dance records. Both he and I have the impression that Nella and her father enjoyed working together in organizing those parties.

I can give an idea of what a good sport Enrico was by relating one ex-perience at those monthly parties. Sometimes between the sets of dancing, there were party games. I proposed a group version of Twenty Questions. I suggested that the guesser be one of the world's best logical thinkers. So Enrico was chosen and he gladly agreed to step out of the room. Then, I proposed to the rest of the crowd that we not choose any object for him to guess, but instead we answer, "yes" if his guess ends in a vowel, "no" if his sentence ends in a consonant and "sometimes yes and sometimes no" if the sentence ends in a y. So we called Enrico back into the room and stood in a circle around him. He could choose anyone in the circle to answer his first yes or no question, and so on. He rather quickly realized that he should ask some redundant questions and then he remarked: "I think you have made up a story with some built-in contradictions." I replied to him: "How could we all come up with the same crazy story and be in complete agreement with each other?" He never did discover the vowel-consonant code and fi-nally had to give up.

Not much later, again by pure coincidence, I encountered Enrico skat-ing by himself at a University ice-skating rink. He greeted me and it seemed only natural to join him. I didn't even give it a second thought. It was clear that he enjoyed young people and we got better acquainted in this and sub-sequent weekly sessions on ice. It was not beneath him to associate freely with students and to treat them as equals. In fact I think he enjoyed young physics students more than some of his older colleagues.

Another example of his enjoyment of young people was that he ate lunch in the large student cafeteria (Hutchinson Commons) rather than the Men's Faculty Club, where most of his fellow faculty members ate. The center long table at the student cafeteria became known informally as the Fermi table; however, anyone was welcome. Several of those who frequented that table later became Nobel Prize winners. In the Chicago physics department of that time, the younger grad students felt that some of the older grad students (like T. D. Lee, C. N. Yang, Geoffrey Chew, Marvin Goldberger, Richard Garwin, Lincoln Wolfenstein, Jack Steinberger, and Marshall Rosenbluth) were better teachers on the whole than the faculty at that time (except, of course, for Fermi, who was clearly the best). Fermi was a modest person, and liked to be treated as one of the crowd. Just to give one example of his modesty, even though one of his many great achievements was the discovery of Fermi statistics, he always referred to it as "Pauli statistics."

❖ ❖ ❖

Arthur Rosenfeld
REMINISCENCES OF FERMI

I am only going to make few reminiscences on some characteristics of Fermi, particularly his modesty and a little bit of his political activism.

I first saw Fermi when I came here as a student in 1946. Fermi would indeed get to his lab at 7:30 a.m. every morning. He would lock his bike to the chain-link fence of the tennis courts in front of Eckhart Hall and would walk in wearing a pretty nondescript parka with a paper bag in his pocket with his lunch, although sometimes he went to the faculty club. What struck me particularly was that the man who ran the stock room also came in about this time. His name was Fred—I can't remember his last name. He had somehow acquired a postwar, black, very impressive Chrysler. Fermi and I on our bikes and Fred would arrive at about the same time. Fred also brought his lunch but thought it would look better in a briefcase. So Fred would get out of his Chrysler and lock it up, and be walking up the steps at about the same time as Fermi. If you were a passerby you would expect this guy in the nondescript parka to dash up the steps and unlock the door, but in fact Fred would dash up the stairs and unlock the door, and we would all go in.

It was in late 1948 that Fermi announced that he was indeed going to give a course on nuclear physics. Jay Orear and Bob Schluter and I were already TAs. We got the idea that we would go see Professor Fermi and ask if he would give us his permission to write up notes. We also managed to be assigned as readers for the course, so we had to solve problems. Professor Fermi was very agreeable to that, and we put up an announcement that we were going to take notes and make them available at the end of each week. I think at the beginning something like seventy-five people had signed up. We started off with the ridiculous idea that we were going to type the notes and reproduce them on a mimeograph machine. Then the requests started pouring in, including from places as far away as Los Alamos. Pretty soon there were three hundred requests, and that was the point where Susie, the department secretary, declared that the notes better get done somewhere else or she was going to quit. As I look back on it, Jay and Bob and I didn't understand the appeal of these notes at the time but Fermi should have known better. Little did we realize that, by the time we got through the fifth or sixth edition, the university press would have sold something like twenty thousand copies.

For those who are statistically inclined, we did go through and make corrections each time through the final edition. We made lots of mistakes and typos. They were never set in type, so we could easily correct them. They were typeset in six other countries, but we are proud of the fact that we made the first camera-ready copy in English. My fit to the number of corrections that we made went down about $1/e$ per edition, so there must still be a few mistakes.

I conclude these brief comments with a recollection that I have of Fermi's last press conference. In fact, it was filmed. It is referred to as the Holton film, with the title *The Times of Enrico Fermi*.

These were the days in 1954 when there was a lot of bitterness. Senator Joseph R. McCarthy was still in power and Robert Oppenheimer was being pictured as a villain. A terrible book came out called *The Hydrogen Bomb* by James Shepley and Clay Blair. I had a habit of walking a few blocks over to the University of Chicago Press at about five o'clock because that was when the *New York Times* arrived in those days. This book had arrived, and I remember buying three or four copies, thinking that it would be pretty interesting. I came back and gave one to Enrico, and I think one to Murray Gell-Mann—I don't remember exactly. I started reading the book, which painted Oppenheimer in a negative fashion: Oppenheimer black and Teller white. I spent most of the night reading the book and came to the lab the next morning. Fermi came in and I asked him what he thought. He said, "Well, I stayed up all night reading that damn book."

I said, "Well, how would you like to have a press conference to tell the world what you think about it?" and he said, "Okay."

So I called the U of C Public Information Office and a friend, Irv Goodman, who worked for *Newsweek*. By the second afternoon we had a very well-attended press conference. Fermi explained that the world isn't black and white. I have the press statement he handed out; it is pretty short, I'll just quote it:

"It is my conviction that the Los Alamos Laboratory does deserve gratitude throughout this nation for the development of both the atomic and hydrogen weapons. This outstanding success is due to the intelligent and self-sacrificing work of the staff and to the sound and foreseeing direction by Norris Bradbury. For this reason I have been deeply perturbed by the implications of this recent book, *The Hydrogen Bomb* by Shepley and Blair, that the laboratory dragged its feet and went only half-heartedly into H-Bomb development. Statements of this kind are bound to produce dissension, and to set back the atomic program. It is true, of course, that Edward Teller is the hero of the H development, but it is equally true that a single man cannot alone carry out a job of this kind. A genius needs the support of many other men in organizations. The Los Alamos Laboratory developed and added to these ideas and brought them into practice."

So I was very pleased with the press conference, and I'm pleased that in the film, in fact, I am sitting in full view, orchestrating the affair.

❖ ❖ ❖

Robert Schluter
THREE REMINISCENCES OF ENRICO FERMI

Fermi's Introductory Physics Text

In 1946 Fermi contracted with the publisher Macmillan to produce an English translation of his Italian "high school" introductory text for the American market. Laura Fermi translated. Macmillan hired a science teacher, Warren M. Davis, principal and science teacher at Alliance High School in Ohio, to review the chapters. Chapters passed back and forth. In the six years that the Fermis gave attention to this project some of the issues arising were

- Fermi emphasizes "principles," but Davis calls for "modern applications" instead.

- Original has arrows over vectors, but Davis feels this is not suitable for American students.
- Original has numbered equations, but Davis believes this would trouble American students.
- Fermi complains that Davis is "overwriting" in his reworking of Fermi's translated passages.
- Also, Fermi states that Davis's proposed "language is suited to younger children."

In the exchange Fermi characteristically enhanced the efficiency and precision of his work by using a code of succinct symbols to convey comments on Davis's changes, which he placed on the returned manuscripts.

"1"   means "grammar fault" (introduced by Davis).
"2"   means "physics is now incorrect—check and correct."
"3"   means "now is generally obscure—rewrite."
"4"   means "why deviate from the translation?"
"5"   means "delete this."
"6"   means "add illustration."
"7"   means "use letters to simplify" (Davis apparently felt that use of letters for quantities was too difficult for U.S. high school students).
"8"   means "delete, or change the question."

On November 10, 1952, Fermi withdrew from the project, writing, "(I) won't sign my name to a book of which I am not entirely satisfied."

During these six years Fermi was president of the American Physical Society; was deeply involved at all levels with the new Navy Large Synchrocyclotron; taught steadily many courses, including the introductory sequence at Chicago, and much else, but nevertheless took time to prepare the book for publication in the United States. Written the way he wanted it to be, it would have been a salutary influence on introductory physics teaching in this country at a time when "dumbing down" was becoming a popular policy.

Notes on Fermi's Nuclear Physics Course

My impetus to transcribe notes from Fermi's Nuclear Physics courses, 262 and 263, January to June of 1949, came from the request of Howard A. Wilcox to send him notes from the upcoming nuclear physics course to be taught by Fermi. Wilcox had finished a Ph.D. with Sam Allison and had joined the Berkeley cyclotron laboratory. I had known Howard Wilcox at

Los Alamos. I am sure that my colleagues Art Rosenfeld and Jay Orear would have wanted to do it in any case.

I asked Fermi if he was agreeable, and he said OK, provided it took none of his time.

We mimeographed our notes on the first three topics of the course and sold 155 copies at our direct cost, $1.50 each. However, mimeographing quickly proved technically inadequate, the stencils tedious, and we approached the University of Chicago press to get the notes "planographed." The result was the white-covered version dated October 1947.

The press got permission from Fermi to produce two thousand in a revised edition dated 1950 (blue cover), incorporating some seventy corrections we were aware of at the time. The preface makes clear that "Dr. Fermi has not read this material; he is not responsible for errors." At this time the press purchased the work from the three of us for $333.33 each.

At this time Fermi emphasized a relatively unsophisticated description of beta decay. The "Fermi Notes" contained a chapter on beta decay which I transcribed. Coming from the source, it had to be as perfect a recording of Fermi's lectures as we could do. Nevertheless, at the 1951 International Conference on High Energy Physics, at Chicago, I was told emphatically by Emilio Segrè that we must be mistaken, it couldn't be what Fermi said, we just misunderstood. At this time Fermi emphasized that the experimental situation in beta decay could be described adequately by a simple form of interaction.

Fermi was unwilling to permit Friedr. Vieweg & Sohn, publishers, to make a formal edition. He told them, "this very able compilation of the notes of my classes by three of my students has been very well received . . . but I haven't carefully read or edited it."

Fermi's Students Take Up Group Theory

In 1950 Fermi invited several students (T. D. Lee, Steve Moszkowski, Jay Orear, Uri Haber-Schaim, and myself—hopefully I haven't forgotten someone—from the first two "basic" qualifying exams) to meet with him an evening each week to study group theory. He said that it seemed to him that anything that could be done using group theory he could do in some way already familiar to him, but group theory formulations were likely to be important, and ought to be mastered. At the first meeting Fermi approached the rotations group. The next time there had been news from Berkeley about the neutral pion, the existence and properties of which were just being glimpsed. From then on, group theory was set aside. Fermi gave exercises (homework) arising from the pion physics news from Berkeley.

❖ ❖ ❖

Jack Steinberger

FERMI AND MY GRADUATE YEARS AT CHICAGO:
HAPPY REMINISCENCES

My two graduate years at the University of Chicago were probably the most satisfying time in my privileged life. Fermi in particular, teacher and model physicist, gave the direction to my subsequent work in physics.

My contact with Fermi was the following:

1946–47	Assistant in Fermi's undergraduate course in physics
1946	Fermi's course on electromagnetic theory
1947	Fermi's course on nuclear physics
1947–48	Evening sessions with Fermi and fellow students on problems in physics
1948	Ph.D. thesis with Fermi
1952–53	Competitor in the measurements of pion-nucleon scattering
1954	Varenna summer school

Fermi devoted a great deal of his time to the graduate students. We were the first group after the war. The department had been completely re-organized by William Zachariasen and was excellent. In addition to Fermi there were Gregor Wentzel, Edward Teller, Maria Mayer, Clarence Zener, and Zachariasen himself. We were a dozen or so students, several (not I) extraordinarily gifted. The war had interrupted everyone's academic life, and we all, faculty and students, were exhilarated by the opportunity and challenge to catch up with lost time, to learn and to do some physics. Fermi taught many courses. His teaching was exemplary, minutely prepared, clear, with emphasis on simplicity and understanding of the basic ideas, rather than generalities and complications. He regularly came to lunch with us, to Hutchinson Commons, the student dining hall. One evening a week, at least during one of these two years, he invited us to a session in which he would propose some problem in physics, unrelated to any course work, and invite us to understand it. The solution would then be discussed the next week. More than once a year we were invited to a dinner in his house. In the parlor games following, whether penny pitching or musical chairs, Fermi liked to win.

Fermi had some particular friends among the students whom he had already known in Los Alamos days, for instance, Joan Hinton, a wonderful girl who, before getting her degree, went on to China and devoted her life to help make this proletarian experiment successful. But whether we were special friends or not, especially gifted or not, Fermi did what he could to help us. We would knock at his office door, and if free, he would take us in, and then he would be ours until the question was resolved.

Fermi agreed to be my thesis adviser, despite the fact that there was no great evidence of capability, and perhaps even evidence to the contrary (I had distinguished myself as the only one to fail the Ph.D. qualifying exam). I still don't know what he thought of my potential as a physicist (I would give a lot to know), but it didn't matter. Fermi was ready to help each one of us, as much as he could. As all my fellow students, I wanted to do a theoretical thesis, but it turned out to be an experiment. In retrospect, perhaps the most interesting fact about Fermi as thesis adviser is that despite his clear pleasure in experimenting, and his interest in the physics of this cosmic ray experiment, he let me do my thing, without participating himself or suggesting the design. He did help me to get things done, such as finding machinists to make the Geiger counters, a lady (Mrs. Woods) to fill them with the proper gas, and a truck with a young man to drive it (I didn't know how to drive), to take the experiment to the top of a mountain in Colorado. But at a certain moment Fermi did give me some advice: when I was analyzing the result, and he saw that I might not do so, he said: "Jack, don't forget to correct for the radiation of the electrons in estimating their energy." This would have been a very bad mistake, which, however, I was capable of making. One incident I enjoy remembering happened on the morning I first turned on the apparatus and found a counting rate which was much too large. I was still shocked and perplexed when the time came to go to lunch, with Fermi and some fellow students: theorists, of course; perhaps it was Murph Goldberger and Geoff Chew. Fermi took pleasure in telling them of my troubles, with the comment: you see, also experiments can have difficulties and challenges. Fermi was of course excellent with both theoretical and experimental problems. Perhaps I understand why he enjoyed experiments so much. Theory can be frustrating; it is not always possible to think of an interesting problem you can solve. Experiments offer more possibility of relaxation: soldering wires, brazing tube joints, or turning things on a lathe.

A few years later I was able to see Fermi, the experimenter, at close range. We were both doing very similar experiments, the measurement of pion-nucleon scattering, Fermi at the Chicago 440-MeV cyclotron, I at the Columbia-Nevis 400-MeV cyclotron. The designs of the experiments—a liquid hydrogen target, liquid scintillation counters, the electronics, and so on—were very similar. But still, watching Fermi at close range gave me several opportunities to again admire his exceptional qualities. One aspect was the clarity of the overall conception of what was interesting to measure, as well as his competence in the theoretical analysis of the experimental data. A completely different cause for admiration was Fermi's invention of an ingenious little cart to move the target inside the cyclotron vacuum, which permitted a change of the pion beam energy without the time-consuming operation of opening the cyclotron vacuum. The cart is shown in figure 7.8. We had been working with this problem for at least a year before Fermi, but this idea had never occurred to us. Also the manner of execution, using the magnetic field of the cyclotron as part of the motor, and the shims of the cyclotron as rails, was beautiful.

I hope that I can capture here some of Fermi's essential personal qualities as a physicist. Fermi was intensely focused on understanding physics. He also cared about family and his own health (he played tennis, came to work on his bike), but had little interest in art. As best I know he did have a real sense of social responsibility as a scientist, in particular on the questions raised by the atom bomb and its evolution, but these questions he did

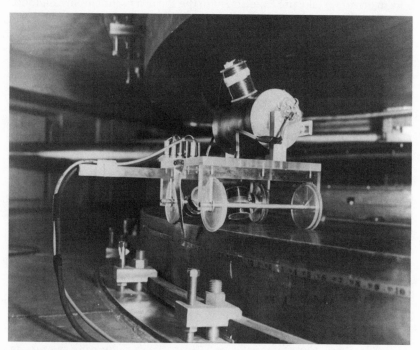

Figure 7.8 Fermi's famous cyclotron trolley. (Photo courtesy archives of the Enrico Fermi Institute, University of Chicago.)

Figure 7.9 An outing on the Grigna near Varenna. *Left to right:* Bianca Puppi, Fermi, Mrs. and Mr. Goldschmidt-Clermont, Eduardo Amaldi, and Jack Steinberger. (Photo courtesy Jack Steinberger.)

not discuss with the students, at least never with me. He had no vanity that I could notice, and needed no reassurance about his greatness in physics. This did not seem to matter to him; what mattered was to understand physics. When a new development of interest came along, he would insist on understanding it thoroughly, deriving the result in his own way, and this work he recorded in notebooks which are preserved to this day. If later he needed to come back to this physics, he knew how to find it in his own notes.

The last time I saw Fermi was at the Varenna conference, in the summer of 1954. Figure 7.9 is a snapshot I took on an outing.

Fermi was outstanding; it was an incredible privilege to have the opportunity at Chicago to learn from and to watch him. But in addition to Fermi there were others who worked together to make the department great, and a privilege to study in. Many, such as Maria Mayer, Edward Teller, Gregor

Wentzel, and Clarence Zener, had been collected as the war came to its end, by William Zachariasen, who had remained from the prewar faculty. Together they reorganized the curriculum. Another privilege, which I am very conscious of, was the contact with the group of fellow students. We worked together in a good spirit, appreciating our great opportunity. Most of my fellow students were well ahead of me in their knowledge of physics and helped me a lot, quite comparable to the help from the professors. In particular, I remember that Frank Yang, coming from China, much of which had been occupied by Japan and which had only one university physics department left during the war (Kunming), already knew essentially all that was taught in our courses.

In the following photographs (figs. 7.10–7.14) are some of the remarkable people who were in the department during my days at Chicago.

Figure 7.10 William Zachariasen, 1948. Zachariasen was chairman of the University of Chicago Department of Physics and later dean of the Physical Sciences Division. He was responsible for rebuilding the department after the war. (Photo courtesy University of Chicago visual archives.)

Figure 7.11 Gregor Wentzel, 1948. Wentzel was an outstanding theoretical physicist and extraordinary teacher. (Photo courtesy Jack Steinberger.)

Figure 7.12 Edward Teller and Lazlo Tisza in 1947. Tisza was not at Chicago but was my highly valued first physics teacher, at MIT, 1943–45. (Photo courtesy Jack Steinberger.)

Figure 7.13 C. N. "Frank" Yang in 1948. (Photo courtesy Jack Steinberger.)

Figure 7.14 Marvin (Murph) Goldberger and Sidney Dancoff in 1949. Dancoff was not our teacher or fellow student, but a valued colleague. (Photo courtesy Jack Steinberger.)

Chapter Eight

Reminiscences of Students
of the Fermi Period, 1945–1954

❖ ❖ ❖

Nina Byers

FERMI AND SZILARD

I. Introduction

I first encountered Enrico Fermi as an undergraduate at
the University of California at Berkeley. I attended his
undergraduate Quantum Mechanics course when he was
a visitor at Berkeley in 1947 and resolved to do all I could
to get into graduate school at the University of Chicago.
Luckily, I was given a teaching assistantship, without
which I could not have afforded to go. A footnote to history here—I was the only female TA then, and heard years
later that some faculty objected to my appointment, because all TAs shared a large office and a female shouldn't
be put there. Nevertheless, I was appointed, presumably
without objection from Fermi. His influence was perva-
sive in the department, so it is unimaginable that an appointment would have been made against his objection.
With me he was cordial, friendly, and encouraging.

Fermi has been a role model for generations of physicists. Though we did not have a comparable gift for phys-

ics, his example of how he did physics and how he enabled us to work enthusiastically and effectively together is truly inspiring. There was another brilliant physicist in Chicago when I was there: Leo Szilard, as different from Fermi as two men could be. But he is also a role model for many of us, sharply contrasting with Fermi, in that he was outspoken about the threat of nuclear war and actively engaged in efforts to avert it. The two men are forever joined in history because together they showed that a self-sustaining nuclear chain reaction could be made to occur. [See the Fermi-Szilard correspondence in chapter 3.—Ed.] Their work began after the discovery of neutron-induced uranium fission in the winter of 1939–40. The first government funds appropriated for the project were designated for studies of the "Fermi-Szilard system." Foreseeing that the federal government would fund physics generously after the war, Robert M. Hutchins appointed them to the faculty of the University of Chicago in 1945, along with other stars of the Manhattan Project. Szilard's appointment was, however, not in the physics department, but in the Institute for Radiobiology and on the Committee on Social Thought. When I was in Chicago, he was working with Aaron Novick in the institute on bacteriophage. He and Fermi appeared not to be on speaking terms. He hired me as something of an intermediary, in 1953–54, to tell him about Fermi's studies of pion-nucleon interactions at the Chicago cyclotron. Fermi was telling us about those in a graduate seminar he gave on pion physics. It was fascinating to learn from Fermi and be exposed to Szilard's questions. They were deep, and I couldn't always answer them, but I was rewarded with a good dinner once or twice a week for expositions from my notes!

For decades afterward I remained curious about why and how these two men became estranged. In my experience both were very civilized, friendly, and cordial human beings. They were very different temperamentally and philosophically. Indeed, Szilard recalled that "On matters scientific or technical there was rarely any disagreement [but] Fermi and I disagreed from the very start of our collaboration about every issue that involved not science but principles of action in the face of the approaching war. If the nation owes us gratitude—and it may not—it does so for having stuck it out together as long as was necessary" [1]. Many who were at Columbia in the early days, including Szilard himself, observed that somehow Fermi found Szilard irritating. Their styles of work were very different.

The history of this collaboration is interesting, particularly so because these two men are starkly contrasting role models of physicists deeply embedded in a moral issue of historic proportion.

II. Some History

These two brilliant, contemporaneous physicists came from hugely disparate cultural backgrounds. Fermi's antecedents were Catholic, Italian farmers. His grandfather became foreman on a large estate and Enrico's father a civil servant working for the Italian railroads. Szilard's family, on the other hand, were Ashkenazi Jews, professionals who had migrated from Eastern Europe to Budapest, where Leo was born. His grandfather was a doctor and his father an engineer and entrepreneur. Both Enrico and Leo were agnostics. Photographs of the two in the mid-1930s, Fermi in Rome, and Szilard in England are shown, respectively, in figure 8.1 and figure 8.2.

Figure 8.1 Enrico Fermi in the mid-1930s in Italy. (Courtesy AIP Emilio Segrè Visual Archives.)

Figure 8.2 Leo Szilard in England, 1936. (Photo courtesy Leo Szilard Papers, Mandeville Special Collections Department, University of California, San Diego.)

Early in 1939 both Fermi and Szilard were staying in the Kings Crown Hotel in New York City and heard from Eugene Wigner the news of the discovery of neutron-induced fission of uranium by Otto Hahn, Lise Meitner, and Fritz Strassmann. Szilard hypothesized that secondary neutrons are emitted in this process and persuaded Fermi to begin experimental study of uranium fission immediately. For Fermi this was a continuation of the neutron work he had begun in Rome five years before. In 1939 he was involved in other work, but Szilard felt they should start this study without delay because uranium might enable construction of a superbomb for use in World War II. Imagining that a self-sustaining neutron chain reaction was possible after the neutron was discovered in 1932, Szilard had been searching for years for a nucleus that would fuel it. He had applied for a patent for the idea in 1934 and wanted to consign it to the British Admiralty on condition they keep it secret. He had been an early anti-Fascist, was in Berlin in 1933 when Hitler took power, and emigrated to England immediately after the Reichstag fire. Fermi and Szilard began experimental studies at Columbia University [2]. Fermi was professor of physics there, and Szilard was given a three-month appointment with laboratory privileges [3].

An early disagreement between Fermi and Szilard was over keeping their work secret. Fermi was opposed to secrecy and censorship. Szilard recalled, "From the beginning the line was drawn. . . . Fermi thought that the conservative thing was to play down the possibility [of a chain reaction], and I thought the conservative thing was to assume that it would happen and take all necessary precautions" [4]. It was not until two years later that Fermi believed a bomb could be built. In the spring of 1941 Mark Oliphant visited Fermi and found he "had no interest in atomic bombs; his aim was to determine whether or not a nuclear pile [reactor] could be made to work" [5]. Oliphant returned from England in the fall to acquaint physicists in the United States with the Frisch-Peierls memorandum [6] that showed an atomic bomb was a practical possibility. In December the Manhattan Project was established under Brigadier General L. R. Groves of the U.S. Army Corps of Engineers. Szilard had prevailed on the secrecy issue in the early days, and publication of results of uranium studies were postponed until after the war.

Army Intelligence was asked to investigate Fermi and Szilard, and their report of August 13, 1940 is reproduced in figure 8.3 [7]. Note that the report ends with "This information has been received from highly reliable sources." It doesn't, however, contain or reflect information the FBI had obtained from interviews with Albert Einstein and Eugene Wigner. The FBI report of the interviews with Einstein and Wigner is reproduced in figure 8.4. As regards the Army Intelligence report about Fermi which states, "He is undoubtedly a Fascist," quite possibly he was a member of the Italian Fascist Party while in Italy under the Mussolini government. To those of us who knew him, however, it is not at all plausible that he believed in or subscribed to Fascist policies or programs. The report does not take into account the circumstances of life in Mussolini's Italy as perhaps it should have done.

III. The Manhattan Project

The project to develop "the Fermi-Szilard system" was taken over by the U.S. Army in September 1942 and put under the jurisdiction of the Manhattan Engineer District, directed by General Groves. The work, however, had moved to Chicago the previous spring, when the chairman of the physics department at the university, A. H. Compton, became its director. It focused mainly on constructing a nuclear reactor (the pile), and as everyone knows, success came on December 2, 1942 under Stagg Field. The group who achieved the first self-sustaining nuclear chain reaction is shown in figure 8.5.

Fermi was given security clearance, but General Groves continued to view Szilard with suspicion and kept him under surveillance throughout

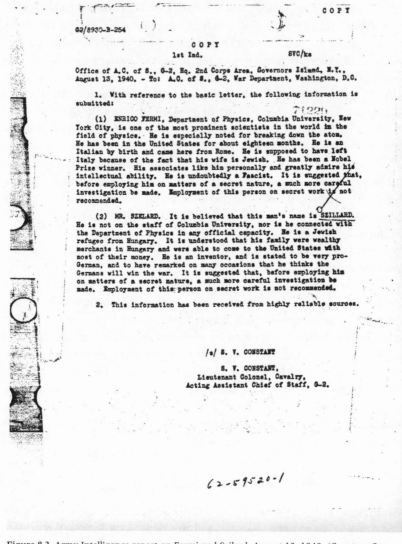

Figure 8.3 Army Intelligence report on Fermi and Szilard, August 13, 1940. (Courtesy Gene Dannen, www.dannen.com.)

the war. The site for the Los Alamos Laboratory was chosen in November 1942 with J. Robert Oppenheimer as director. Szilard never received security clearance to go there. General Groves had what appears to be an irrational distrust of Szilard's loyalty. This is expressed, for example, in a memo dated June 12, 1943 [7]:

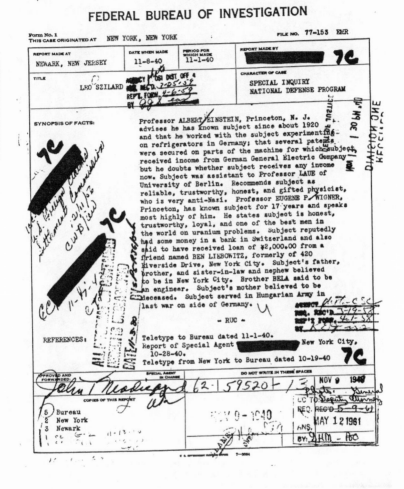

Figure 8.4 FBI interview report on Leo Szilard, November 8, 1940 (Courtesy Gene Dannen, www.dannen.com.)

Attention: Captain Calvert

1. Reference your letter of June 9, 1943, MI-1a, subject is above, the investigation of Szilard should be continued *despite the barreness* [*sic*] *of the results* [italics mine]. One letter or phone call once in three months would be sufficient for the passing of vital information and until we know

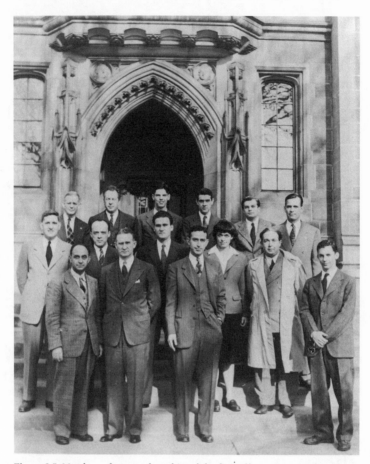

Figure 8.5 Members of group who achieved the first self-sustaining nuclear reaction on December 2, 1942. The photograph was taken on December 2, 1946. *Left to right, first row:* Fermi, Walter Zinn, Albert Wattenberg, and Herbert Anderson. *Middle row:* Harold Agnew, William Sturm, Harold Lichtenberger, Leona Woods Marshall, and Leo Szilard. *Back row:* Norman Hilbery, Samuel Allison, Thomas Brill, Robert Nobles, Warren Nyer, and Marvin Wilkening. (Photo courtesy Argonne National Laboratory.)

for certain that he is 100% reliable we cannot entirely disregard this person.

L. R. Groves
Brigadier General, C.E.

Szilard was bemused by the continuing surveillance. He liked to tell about sometimes shaking the agents following him off of his trail and then greeting them and, on a cold winter's day in Chicago, inviting them for a hot cup of coffee. He was a kind and good-humored man, and had a sense of proportion. He took very seriously the threat of fascism and nuclear war, but not so seriously Groves's distrust of him.

IV. 1943–1945

After Oliphant's second visit in 1941 attention shifted from developing reactors for nuclear power to nuclear weapons development. The plutonium isotope Pu^{239}, fissionable like U^{235}, was discovered at the Berkeley cyclotron by Glenn T. Seaborg and collaborators. It is made in relatively large quantities in the graphite-uranium reactor Fermi and Szilard had designed and built. They worked on improved versions which were built in the Argonne Forest, a suburb of Chicago. Seaborg brought a group of chemists to Chicago to extract the plutonium. The project at the university became known as the Metallurgical Laboratory, the Met Lab. In September 1944 Fermi went to work with Oppenheimer in Los Alamos. Close collaboration with Szilard had ended earlier.

In the winter of 1943–44 the tide of war turned in the Allies' favor. Hitler's armies were in retreat in Russia, Allied bombing was causing massive damage in Germany, and intelligence sources reported no evidence of a German bomb project. A German bomb no longer seemed an imminent threat; presumably, if the Germans had an atomic bomb, they would have used it. Szilard and others began to worry about consequences of military use of nuclear power. Niels Bohr, in an effort to avoid a nuclear arms race and promote international control, personally urged President Roosevelt and Prime Minister Churchill to share with all nations the nuclear technology [8]. His suggestion was memorialized and decisively rejected by Roosevelt and Churchill in their 1944 Hyde Park Aide-memoire [9.26]. "Met Lab scientists and engineers recognized early what the impact of the release of nuclear energy would mean for the future of society and grappled with the question," John Simpson recalled; despite security restrictions, they were able to hold seminars and discussions around this issue [10]. Szilard, Seaborg, and Nobel laureate James Franck were among those actively engaged.

Apparently dropping the bomb in Europe was never systematically considered. Historians find the common view in the military that if and when it was ready it would be used against Japanese forces. For example, minutes of the Military Policy Committee meeting of May 5, 1943, read, "The point of use of the first bomb was discussed and the general view appeared to be that its best point of use would be on a Japanese fleet concentration. . . . The Japanese were selected as they would not be so apt to secure knowledge from it as would the Germans" [11].

In March 1945, with old friend Einstein's help, Szilard arranged an appointment with President Roosevelt to argue that the long-term danger of using the weapon outweighed the short-term military advantage its use would bring. He prepared a memo for this meeting which is remarkably prescient [12]. He warns that relatively small nuclear bombs could be made in the near future and says, "it seems the position of the United States in the world may be adversely affected by their existence. . . . Clearly, if such bombs are available, it is not necessary to bomb our cities from the air in order to destroy them. All that is necessary is to place a comparatively small number of such bombs in each of our major cities and to detonate them at some later time. The United States has a very long coastline which will make it possible to smuggle in such bombs in peacetime and to carry them by truck into our cities. The long coastline, the structure of our society, and our very heterogeneous population may make effective control of such 'traffic' virtually impossible." Roosevelt died on April 12, before the meeting could take place. As vice-president, the new president, Harry S. Truman, had not known about the Manhattan Project. In late April, at the urging of Vannevar Bush and James Conant, he appointed a committee to advise the president and Congress on issues relating to nuclear energy. It was called the Interim Committee [13]. In the Met Lab, a committee called the Committee on Political and Social Problems of the Metallurgical Laboratory wrote a report for the Interim Committee. It has come to be known as the Franck Committee after its chair, James A. Franck [14]. Its report, now famous as the Franck Report, was classified Top Secret and not made public until years after the end of World War II [7, 9.49, 12].

A reading of the Franck Report indicates the committee did not know that the Interim Committee, in its June 1 meeting, had unanimously recommended that "the bomb should be used against Japan as soon as possible; that it be used on a war plant surrounded by workers' homes, and that it be used without prior warning" [9.44]. This is a decision the Franck Committee had hoped to forestall (see below). It was recommended by James Byrnes, secretary of state designate, an Interim Committee member. According to Richard Rhodes [4], President Truman adopted the Interim Committee's recommendation after a meeting with Byrnes later that day (June 1). His official decision, however, was not announced until after the bomb had tested successfully at Alamogordo.

The Franck Report is a lengthy, deliberative document beginning with something of an apologia for presuming to advise the government: "we found ourselves, by the force of events, the last five years in the position of a small group of citizens cognizant of a grave danger for the safety of this country as well as for the future of all the other nations, of which the rest of mankind is unaware. We therefore felt it our duty to urge that the political problems, arising from the mastering of atomic power, be recognized in all their gravity, and that appropriate steps be taken for their study and the preparation of necessary decisions. We hope that the creation of the Committee by the Secretary of War to deal with all aspects of nucleonics, indicates that these implications have been recognized by the government. We feel that our acquaintance with the scientific elements of the situation and prolonged preoccupation with its world-wide political implications, imposes on us the obligation to offer to the Committee some suggestions as to the possible solution of these grave problems." It concludes with, "We believe that these considerations make the use of nuclear bombs for an early, unannounced attack against Japan inadvisable. [This would] precipitate the race of armaments and prejudice the possibility of reaching an international agreement on the future control of such weapons. Much more favorable conditions for the eventual achievement of such an agreement could be created if nuclear bombs were first revealed to the world by a demonstration in an appropriately selected uninhabited area. . . . To sum up, we urge that the use of nuclear bombs in this war be considered as a problem of long-range national policy rather than military expediency, and that this policy be directed primarily to the achievement of an agreement permitting an effective international control of the means of nuclear warfare."

A. H. Compton, director of the Met Lab, transmitted the Franck Report to Secretary of War Stimson with a covering letter (dated June 12) that expresses his lack of concurrence with it [9.48]. It seems likely he knew that the Interim Committee had recommended immediate use of the bomb. His brother Karl was on that committee. Probably to "balance out" the Franck Committee's contrary recommendations, he referred its report to the Interim Committee's advisory Scientists Panel consisting of himself, J. Robert Oppenheimer, Ernest O. Lawrence, and Enrico Fermi. Their report (dated June 16) supports the Interim Committee's June 1 recommendation [9.51].

The following observation of Szilard sheds some light on why it was that Fermi and Szilard gave such different advice: "Fermi is a scientist pure and simple. This position is unassailable because it is all of one piece. I doubt he understood some people live in two worlds like I do. A world, and science is a part of this one, in which we have to predict what is going to happen, and another world in which we try to forget these predictions in order to be able to fight for what we would want to happen." These men were caught in most distressing historical circumstances. They must have agonized deeply about the consequences of their work. Fermi did not express himself on these matters, and we do not know what he was thinking [15]. Whatever it might have been, it seems unlikely he would have taken exception publicly to the conclusion expressed by the others on the Scientists Panel.

It is a matter of debate among historians whether or not the atomic bombing of Japanese cities was needed to obtain surrender before the planned November 1 invasion of the home islands [16]. Nobel laureate P. M. S. Blackett draws attention to the August 6 and 9 timing of the drops given that it was known the Soviet Union would enter the war against Japan no later than August 8 [17]. He suggests that political rather than military advantage motivated the planners—the age-old game of power politics. Certainly the atomic bombing on August 6, two days before the Soviet invasion of Manchuria and Korea, effectively foreclosed any possibility of the Soviet Union's participation with them as victor in the Japanese war. This would not be the first time military power was used for political advantage, but it was the first time nuclear power was used in this way. From the vantage point of more than fifty years hence, it would seem to me to have set a dangerous precedent.

A moral dilemma faced the Scientists Panel and the scientists of the Manhattan Project more generally. The Franck Report, and a petition to the president [7] based on it, represents the response of people in the Met Lab and Oak Ridge. Most people in Los Alamos were unaware of the report, and of the petition. R. E. Peierls wrote in his autobiography, "In Los Alamos we had no immediate contact with the American authorities, but we knew that Oppenheimer was in contact with them, and we had confidence both in his understanding and his talent for clear exposition. We felt the leaders were reasonable and intelligent people, and would make responsible decisions. In retrospect it is clear these views were too optimistic. I do not wish to imply that the decision makers were lacking in good will, but we overestimated their vision and their ability to adjust to a drastically new situation" [18]. Scientists tend to be more foresighted than government leaders, and their advice has generally been disregarded when not in support of government policy. To Fermi and Szilard and others, it must have seemed in 1945 inexorable that the weapons, which had been constructed at such great expense, would be used.

V. Conclusion

In conclusion, I would like to add a few personal thoughts. Though I have all my adult life opposed the construction and use of nuclear weapons and refused to work on associated projects, I feel humbled by the good fortune I have had to have known and been taught by Enrico Fermi. Had I had the choice I would like to have been an activist along with Szilard and Bohr in 1944–45. After the war in Chicago, many of us actively opposed the testing and construction of nuclear weapons. As far as I know, Fermi did not express disapproval of what we were doing. Since he did not express himself to us on these issues, we do not know what he thought or felt about them. One cannot but believe that as a scientist and teacher he was a deeply humane man. He must have had great understanding and empathy for others. He could not have been such a great teacher without those qualities. He so very generously shared his intellect, creativity, and knowledge with colleagues and students that those of us who knew him can only be grateful for the riches he brought us. It is not for me to judge him as regards his neutrality in the great political issues of his time. Paraphrasing Emilio Segrè, I'd like to end by saying—presidents and wars come and go but Fermi-Dirac statistics is forever.

Acknowledgment

Gene Dannen has amassed an extensive archive of information and documents relating to the life and work of Leo Szilard and made some of this available on the Internet at http://www.dannen.com/szilard.html. I want to thank him for generously sharing knowledge and documentary material with us for presentation in this essay.

Notes and References

1. William Lanouette with Bela Silard, *Genius in the Shadows* (Chicago: University of Chicago Press, 1992).

2. Herbert Anderson, Fermi's graduate student, measured uranium absorption cross sections, and Szilard did likewise with Walter Zinn.

3. Szilard never held a teaching position.

4. Richard Rhodes, *The Making of the Atomic Bomb* (New York: Simon and Schuster, 1986).

5. R. H. Dalitz, private communication; also Oliphant obituary in *Physics Today*, July 2001, p. 73.

6. O. R. Frisch and R. E. Peierls considered fast-neutron-induced fission of U^{235} and showed a bomb could be made with only about 20 kg of U^{235}.

7. Courtesy Gene Dannen. See http://www.dannen.com/szilard.html. This is a very rich Web site containing numerous historical documents obtained from government files under the Freedom of Information Act.

8. M. Gowing, *Britain and Atomic Energy 1939–1945* (London: Macmillan, 1964).

9. M. B. Stoff, J. F. Fanton, and R. H. Williams, eds., *The Manhattan Project: A Documentary Introduction to the Atomic Age* (Philadelphia: Temple University Press, 1991). There are ninety-five numbered documents reproduced in this book. For example, the Churchill-Roosevelt Aide-de-memoire is document 26 and we reference it as 9.26. The Franck Report is document 49 and we reference it as 9.49. Other documents to be found in this volume are similarly referenced.

10. John A. Simpson, "Some Personal Notes," *Bulletin of the Atomic Scientists* 37, no. 1 (January 1981): 26. Simpson was a young physicist on the project at the time and later became the first chairman of the Atomic Scientists of Chicago, a precursor of the Federation of Atomic Scientists, now known as the Federation of American Scientists (FAS).

11. Martin J. Sherwin, *A World Destroyed* (New York: Alfred A. Knopf, 1975). For details see footnote on p. 209 of the first Vintage Books edition (New York: Vintage Books, 1976).

12. See, e.g., "The Atomic Age: Scientists in National and World Affairs," in *Articles from the Bulletin of the Atomic Scientists, 1945–1962,* ed. M. Grodzins and E. Rabinowitch (New York: Basic Books, 1963) (chapter 1 is the Szilard memo; chapter 2 is the Franck Report).

13. The committee members were Secretary of War Stimson (chair), Vannevar Bush, James Conant, Karl Compton (president of MIT and brother of A. H. Compton), Assistant Secretary of State William Clayton, Undersecretary of the Navy Ralph Bard, and Secretary of State Designate James F. Byrnes. It was called interim because Congress had not yet been told about the Manhattan Project.

14. Its members were Nobel laureate James A. Franck, T. R. Hogness, Donald J. Hughes, C. J. Nickson, Eugene S. Rabinowich, Glen T. Seaborg, J. S. Stearns, and Leo Szilard.

15. Laura Fermi, *Atoms in the Family* (Chicago: University of Chicago Press, 1954).

16. See, e.g., Philip Nobile, ed., *Judgment at the Smithsonian* (New York: Marlow, 1995); Barton J. Bernstein, ed., *The Atomic Bomb: The Critical Issues* (Boston: Little, Brown, 1976); "Strategic Bombing Survey Report"—document 94 in reference 8; and P. M. S. Blackett, "The Decision to Use the Bomb," in *Fear, War, and the Bomb* (New York: McGraw-Hill, 1949).

17. The Yalta agreement, signed February 11, 1945, by Churchill, Roosevelt, and Stalin, specified that the Soviet Union would enter the war against Japan no later than August 8, 1945. The Hiroshima bomb was dropped August 6, and the Soviet armies invaded Manchuria and Korea on August 8.

18. R. E. Peierls, *Bird of Passage* (Princeton, N.J.: Princeton University Press, 1985).

❖ ❖ ❖

Jerome I. Friedman
A STUDENT'S VIEW OF FERMI

In 1950, I entered the physics department of the University of Chicago after completing my work in the highly innovative and intellectually stimulating college established by Robert Maynard Hutchins, who was then president of the university.

The physics department was a whole new world, one that was equally stimulating and demanding. The unquestioned intellectual leader of that world was Enrico Fermi. It is difficult to convey the sense of excitement that pervaded the department at that time.

A number of factors combined to create this especially lively atmosphere. Among these were Fermi's brilliance and leadership and the absolutely outstanding physics faculty that had been assembled at Chicago. The many notable physicists who came to visit Fermi also contributed to the excitement in the department. In addition, the pioneering investigations of pion-proton scattering at the newly constructed synchrocyclotron gave Chicago a leadership position in experimental particle research. These all made the department a very special place at a special time in the development of particle physics.

My first recollection of Fermi is of a talk he gave to incoming students. I vividly remember his advice to us to become experimentalists because the time was especially ripe for great progress in experimental physics, and field theory had come to a dead end. Many of us had wanted to be theorists; but we changed our minds as a result of this advice from someone whom we so greatly respected and admired. The ratio of theoretical to experimental students in our class was very small.

My next contact with Fermi occurred when I was a student in his famous courses in thermodynamics and statistical mechanics which were given in the autumn of 1951 and the spring of 1952. These were extraordinary courses, covering in addition to classical physics many aspects of atomic physics, nuclear physics, condensed matter physics, and astrophysics. It was clear to us that Fermi had enormous breadth, as well as depth, in his knowledge of physics. He encompassed all of physics.

In the winter and spring of 1953, he gave courses in nuclear and particle physics and covered virtually all that was known in these two areas, which were closely related at that time. His presentations were the ultimate in clarity and his legendary physical approach to demystifying problems often gave students the impression that the results were obvious. But they were not so obvious when we tried to reproduce them on our own, because we didn't have Fermi's powerful physical insight and his ability to simplify complex problems. Nevertheless, we all tried to emulate Fermi's approach to solving problems.

When I passed the Ph.D. examination—called the basic exam—I summoned up the courage to ask Fermi if I could do my doctoral research under his supervision. To my great surprise, he said yes; and I was overjoyed to be given this wonderful opportunity. The emulsion lab that Fermi had established became the center of my activities. The lab's day-to-day operations were looked after by Art Rosenfeld, and some of the other people who were working in the lab at the time were Jay Orear, Gaurang Yodh, Elliot Silverstein, Bob Swanson, and Horace Taft.

Fermi was highly solicitous with regard to having his students understand physics. With patience and good humor he was invariably willing to explain any aspect of physics that had escaped their understanding, and he was readily available. As I recall, when his office door was open, which was often, one could always go in to see him. Only when it was closed was he unavailable.

On a few occasions he invited me to social gatherings at his home. He and his elegant wife, Laura, could not have been more hospitable and gracious. He personally took me, a very junior graduate student, around the room to meet his guests. This was another measure of his regard for his students. I noticed that at about 10:30 in the evening all of the guests almost simultaneously left. I was told that Fermi went to bed early and all of his guests were aware of this. He was meticulous about time. Speakers at seminars were told that they would have to finish by 6 p.m. if they wanted Fermi to be present for their entire presentation. He always promptly left at that time and arrived at work early in the morning.

I was an occasional helper on his cyclotron runs. During one run there was something that had to be made in the machine shop, and I offered to do it for him. But it was clear that he wanted to do it himself. This man of genius did not consider working in the machine shop beneath him, and he actually seemed to enjoy it. It was my impression that he took a special pleasure in the mechanical devices he designed and made. One of these was the internal target mover that he made for the synchrocyclotron, often referred to as Fermi's trolley. And another was a mechanical analogue of strong focusing that he gleefully demonstrated at a colloquium that he gave.

Fermi's presence at Chicago not only attracted an outstanding faculty, but also drew some of the most renowned physicists in the world as visitors to Chicago. I clearly remember Wolfgang Pauli's and Werner Heisenberg's visits and the excitement their presence caused among the students.

Up-and-coming young physicists were also attracted. For example, Richard Feynman visited a few times to give Fermi private lectures about his latest calculations, and the people working in Fermi's lab were invited. I can recall Fermi listening with a smile on his face as Feynman, speaking with a somewhat exaggerated Brooklyn accent and great animation, reported his latest results on liquid helium.

We had many illustrious speakers at our seminars and colloquia. Fermi inevitably would have penetrating comments and questions. His questions were gentle but sometimes devastating to the speaker, and they usually started off with the phrase "There is something that I do not understand." After seminars and colloquia, Fermi and Edward Teller often had fascinating exchanges, which sometimes were very much like sparring matches. I often felt that I was observing the discourse of titans who spoke in a language that I did not yet fully understand.

For my thesis research, Fermi suggested that I carry out a nuclear emulsion investigation of proton polarization produced by nuclear scattering, an effect which had been observed at cyclotron energies. The objective of this study was to determine whether the polarization resulted from elastic or inelastic scattering. I did not know at the time that Fermi had already theoretically shown that elastic nuclear scattering could produce large polarizations. This calculation was in his famous notebook of problems that he had investigated and solved. The calculation was, as usual, based on a simple model, utilizing a real and an imaginary nuclear potential and a spin-orbit coupling term. This is the same term that he had suggested to Maria Mayer as possibly playing a role in the structure of the nucleus and which was crucial to her development of the shell model.

When I had only partially completed scanning my emulsion plates, Emilio Segrè visited Fermi and told him that he had observed large polarizations in nuclear elastic scattering in a counter experiment at the Berkeley cyclotron. According to Segrè, on the morning of his visit, Fermi calculated the polarization produced in elastic scattering, and his results matched Segrè's measurements beautifully.

I had been scooped and was quite dejected. However, Fermi was very understanding and suggested that I continue my measurements. First, it would be valuable to confirm Segrè's results with another technique; and second, I could also determine to what extent inelastic scattering produced polarization.

During this period I also sat in on Fermi's wonderful courses in quantum mechanics, which were given in the winter and spring of 1954. These were the last courses he gave at Chicago. That summer he went to Italy, where he became ill. When he returned to Chicago in September, I saw him in a corridor at the institute. We were some distance apart and we waved to one another. I was struck by how gaunt he looked. The next day he underwent exploratory surgery at Billings Hospital and was found to have inoperable cancer. I never saw him again.

Subrahmanyan Chandrasekhar told me that when he and Herb Anderson first went to visit him at the hospital, they were initially at a loss for words.

Fermi sensed this and put them at ease by asking, "Tell me Chandra, when I die will I come back as an elephant?" After that remark, the conversation proceeded smoothly. Fermi was truly a remarkable man in all respects.

Fermi died on November 29, 1954. It was a terrible loss not only for our department but also for the world of physics. In 1955, I completed the project that Fermi had assigned to me, verifying Segrè's results with different nuclei and showing that inelastic scattering had no measurable polarization at cyclotron energies. John Marshall kindly signed my thesis.

After Fermi's death, I was asked to gather up his books and journals for the University of Chicago. I don't remember finding any journals, but I found about four books. I no longer remember the titles or authors of the books, except for one which was probably there for sentimental reasons. It was written by his boyhood friend, Enrico Persico. It was clear that Fermi didn't need books. He worked out everything himself.

I was indeed fortunate to have been taught and supervised by this giant of physics and to have seen the practice of physics carried out at its very best, at such an early stage in my development. Nearly half a century later, I still look back in awe at this great physicist and remarkable human being.

❖ ❖ ❖

Maurice Glicksman
ENRICO FERMI: TEACHER, COLLEAGUE, MENTOR

I was an engineering physics undergraduate at Queen's University in Canada, excited by "nuclear physics." I decided to head for where the action was, took the graduate entrance exams, and came to Chicago in 1949 with a scholarship and a desire to work with Enrico Fermi. Right away Dick Garwin invited me to join his laboratory. But that lasted just the first year; Dick decided to go elsewhere, and Herb Anderson approached me to join his group.

I was an avid course taker those first two years. Enrico Fermi and Gregor Wentzel stood out for me as teachers, although very different in their approach. Wentzel was precise and detailed and rarely referred to notes during his lectures. The equations of advanced quantum mechanics were written on the board in an order easy to follow, along with a clear explanation step by step. Problems were easy to solve if you paid attention in class, since they required the same kind of approach as Professor Wentzel demonstrated in class.

Fermi's interest was in having the students understand the physics deeply—as he did—and his reasoning was clearly presented and well ordered. But the course was incomplete without doing the problems which he supplied. He did not solve them in class. They tested your understanding of the physics and were difficult but rewarding. I recall one course in which the advanced graduate student graders asked us students to work with them to understand and solve the more difficult exercises.

One of the courses I took with Enrico was a graduate course in solid-state physics, to be team taught by Fermi—who would provide the theory—and a metallurgist/physicist colleague—who would discuss experiments. The first week Fermi lectured, bringing atoms closer together to give the interaction which led to condensation and solidification as the motion was subdued. The second week, his colleague lectured. Enrico attended the first day with notebook and sharpened pencil, sitting in the front row. But the lecturer led us through alloy phase diagrams, one after the other, and Enrico did not write anything down. He also did not come to the other lectures that week. The rest of the course was taught only by Fermi, who covered experiment as well as theory.

At that time Fermi was interested in understanding the interaction of π-mesons with protons, and the work on the synchrocyclotron was devoted heavily to scattering experiments. He and colleagues Herb Anderson and Darragh Nagle, together with graduate students Ron Martin and Gaurang Yodh, had measured total cross sections and angular distributions of the scattering at the low and intermediate π-meson energies, up to 135 MeV. Particle beam intensities were low, and hence the results had large statistical errors. Fermi was interested in greater precision in order to learn more about the interactions and suggested that I build a cloud chamber to look in detail at the scattering processes. I went off to the University of Michigan to learn about cloud chambers from a summer course taught by Carl Anderson.

But I had grander dreams. The same cloud chamber could also look for rare events, including possible new particles created at the highest energies. Hence, I worked with the synchrocyclotron to extract higher-energy π-meson beams, hoping to use the cloud chamber to find some exciting new particles as well as do the experiments Enrico wanted done. I had then to convince Herb Anderson that this was a feasible proposal, so that he could provide the funds to support the construction.

My visit with Herb in his hospital room was crucial for my proposal. He rejected my plans for a cloud chamber, saying that it would take too long (five years and not the eighteen to thirty-six months I had "naively" estimated) and that his conscience would not let a married graduate student with a new baby starve that long. He encouraged me to search for the possible new particles, using the scintillation counter technology the group had already developed, but also to do π-meson scattering experiments at high energies. The latter would provide useful new data which would be essential for me to get a Ph.D. if I did *not* find new particles.

Fermi's response to this plan—aside from his disappointment in my decision to work at higher energies only—was to insist that my angular distribution measurements be taken at more angles than the group's earlier work, because of the possible greater variation with angle due to higher-order terms.

The results of my first set of experiments were a surprise. I saw many more events than we had expected, indicating a much increased interaction cross section at higher energies. My excitement was shared by others in the group, and Herb approached me to tell me of Fermi's interest in joining with me to do total cross section measurements at higher energies, as an extension of the group's earlier work. My reluctance to share credit for the exciting new results was overcome by Herb's advice that I would never be sorry to author a paper jointly with Fermi.

Enrico insisted on being a full partner, working in the control room, taking data and using his slide rule to do a running calculation of the cross sections as the data appeared. He did have one rule he told me he had adopted with his advanced age: he stopped working in the lab at about 6 p.m. I also observed his aversion to being drawn into political matters. I believe it was Linus Pauling who was trying to get him to support a particular plan of his. During one of our experimental runs, Enrico refused to take Pauling's call, as his secretary asked him to do, insisting that he could not be interrupted while doing his research and telling me he was happy to have the excuse. [Note, however, Fermi's letters concerning Pauling in chapter 4.—Ed.]

The paper we worked on turned out to have some long-term significance, showing an apparent peak as a function of energy in the total cross section for π-meson-proton scattering. My thesis measurements of the angular distributions of the π^--meson scattering provided strong evidence that this was a resonance relating to a new particle, which we called the spin 3/2-isotopic spin 3/2 resonance at that time.

In 1953, as the experiments continued, Enrico spent time at Los Alamos analyzing the data that we had gathered in order to determine the underlying interactions involved. Working with Nick Metropolis and E. F. Alei, he came up with a solution which showed no resonant behavior in any of the scattering phase shifts.

I took two cracks at independently analyzing our data. The first one was fairly simple: I set three of the six phase shifts to zero, following a suggestion of Gregor Wentzel, and using a graphical method found several solutions, one indeed showing a 3/2-3/2 resonance. (Ron Martin, then at Cornell University, had independently found the same result, using a similar approach.) The second involved a computer solution, using the Argonne computer, of a least-squares analysis involving three phase shifts and the equations describing the π^--meson scattering. This also showed a 3/2-3/2 resonant solution to be the most likely.

Herb Anderson was ailing, so that Enrico Fermi chaired my thesis committee. We discussed two matters on which we did not initially agree. One had to do with how to treat possible systematic errors in the data. In their earlier experiments, Enrico and his colleagues had made a number of adjustments to the data, for absorption in the counters and other material which intercepted the particles being studied, gamma ray conversion values, and so on. After all of these corrections, he felt that there were potential systematic errors which were not calculable and insisted that an uncertainty of 10% be included. In those earlier experiments, many of the counts were not large, so that the statistical uncertainty was usually larger than 10%, and this additional factor did not materially increase the reported uncertainty range. My own experiments involved longer counts, and the statistical and other uncertainties were usually about 5%. Enrico insisted on my adding the 10%. He felt that the conservative approach he took here had been amply justified over the years of his experience.

The other matter was the treatment of the phase shift analysis of the angular distribution of the scattering. There were several different sets of solutions, and the data did not definitively distinguish between them, although the resonance solution which I had found was about twice as probable as the other. Enrico cautioned me to be careful in stating my conclusions. As best I can recall, he said: "Maurice, theorists do not understand experiments. We experimentalists have to be careful in explaining our results. The theorists will read your paper, take your conclusions as fact, and retire to the mountaintops to try to explain your results. When they come back down a year later with their new theories and discover that new experiments have shown different results, your name will be MUD, M—U—D."

I listened, and I softened my conclusions, trying to make clear that they were not definitive. My solution showing a resonance was confirmed in more detailed later measurements of π^-- and π^+-meson scattering at a number of energies above and below the resonance. But his advice was good, and I have taken it to heart on many occasions.

Enrico Fermi chaired my thesis defense. There were good questions on the research which I was able to handle satisfactorily, but one of the faculty asked me a related physics question and refused to accept my response as correct. Enrico said that if he thought my answer was incorrect or incomplete, he should give us the correct answer. His response was that he was there to ask questions, not to answer them! The impasse was resolved by Fermi, who proceeded to restate my answer in a different fashion—and this had to be accepted.

Parties at the Fermi house were good fun. But Enrico had a rule that he applied when the guests were a mixed group of physicists and nonscientists: there was to be no physics shoptalk. One evening Ray Bowers and I were engrossed in a "physics" discussion in the living room when we were interrupted by Enrico, who took us each by an arm and moved us into the empty kitchen. He suggested we continue our discussion there by ourselves, and rejoin the party when we were ready to follow the rule.

At times Fermi showed his admiration for the skills of others. One Friday night Herb Anderson was treating our graduate student group to dinner, but Herb showed up at the restaurant quite late. In making his apology, Herb (who was then directing the Institute for Nuclear Studies) told us that he had been waiting to consult with Enrico but had not wanted to interrupt the discussion Enrico was having with that day's guest, John von Neumann. Herb recounted that when he finally saw Enrico, Fermi started off by saying how impressive von Neumann was:

"You know, Herb, Johnny can do calculations in his head ten times as fast as I can!"

Enrico continued, with some humor, "And I can do them ten times as fast as you can, Herb, so you can see how impressive Johnny is!"

Since Herb was no slouch at figuring things in his head, we were properly impressed.

One late December day we were all working in Enrico's office, looking at some of our data and those coming from other laboratories, when someone came in to remind us that a Christmas party was going on in the institute and Fermi should make an appearance. When we arrived, Enrico was greeted by his secretary, who had already enjoyed a few drinks. She came right up to him and said, "Professor Fermi, I was at a party last night and told everyone I worked for the greatest scientist in the world. Doesn't that deserve a kiss?"

She put her arms around him, gave him a big kiss and held on. Fermi stood there blushing, his arms down at his sides. He turned to Nathan (Nate) Sugarman, who was reputed to have a way with women, and said, "Nate, please help me! You know how to handle this." Nate came to his rescue.

The summer of 1954, Enrico Fermi was in Europe, lecturing at Varenna in Italy. A number of us were working in his office-conference room on the day he returned to Chicago. When he came in, our joy at his return was tempered by our shock at his gaunt and pale appearance. He told us that he had not been feeling well for a while, but had waited for his return to the United States to consult a physician. I recall his comment that he had been unable to eat. Turning to me he said that he wished he could enjoy his food like I did, but that everything he put in his mouth tasted like "mud." We urged him to get to a doctor as soon as possible.

Several months later I left Chicago. It was on November 29, 1954, and I knew that Enrico Fermi had died that day. A great light in my life, and that of many others, went out that day.

❖ ❖ ❖

Marshall N. Rosenbluth

A YOUNG MAN ENCOUNTERS ENRICO FERMI

Toward the end of World War II, as I was pondering my postnavy path in life, my father sent me a clipping from a Chicago newspaper enumerating all the physics luminaries slated to teach at Chicago after the war. I didn't know much modern physics but I did know the name Fermi, so I applied for admission.

In fact, my first physics exposure at the University of Chicago was to Fermi's famous 8:00 a.m. lectures on nuclear physics. By about 8:15 I knew I had made the right career decision! I was of course overwhelmed by the way he could reduce the most complex problems to their essence and make the matter both comprehensible and beautiful. In any event, midway through the first quarter, I detected a terrible blunder in a scattering theory lecture—dividing by zero or neglecting some divergence—so, with some trepidation, I knocked on the great man's door and announced my objection. Without a word he went over to his vast set of filing cabinets and pulled out the appropriate file for me to look at. Of course, it was a derivation of the same result done with great elegance and mathematical precision. When I returned the file and agreed I was now convinced, he said with a twinkle in his eye "Please, young man, don't ever tell anyone that I do this kind of thing, or I will lose my credentials as an experimentalist."

While I was not a Fermi student, I think it is accurate to say that his aura and way of looking at physics dominated the intellectual ambiance in the physics department. Much of this I absorbed from interactions with my wonderful fellow graduate students. I did of course have the opportunity at least to participate in many lunches and group discussions with Fermi. I recall vividly that at one lunch Fermi brought up the topic of extraterrestrial civilizations. Running through the probable number of equivalent solar systems, and the factor of a million between the earth's age and the existence of civilization on earth, he pointed out that it was extremely strange we didn't have convincing evidence of much more advanced "alien contact." This much of his thinking has been recounted many times. However, I have never seen his grim conclusion retold. Trying to work out the probability of the many possible explanations of this strange situation, Fermi said that the only one that seemed plausible to him was that when civilization reached the point, as we just had, that it was able to destroy itself completely—it inevitably had done so. Let us hope that at least in this one instance Fermi was wrong!

Just as my beginnings at Chicago were related to Fermi, so was my final academic endeavor—my thesis defense. I was about five minutes into my presentation—an already outdated attempt to formalize a compound π-μ meson theory—when a crackling argument erupted between Fermi and my supervisor, Edward Teller. This quickly wandered far astray from my prosaic offerings and turned into a brilliant and awe-inspiring discussion of the state of modern particle physics. I think my thesis defense must be one of the most intellectually penetrating in history, even if I contributed almost nothing to it! As the hour quickly ended, Fermi turned to Teller and, in his quiet but forceful way, said, "Well, I guess you pass, Edward—and you too, Marshall. Congratulations!" I was gratified, if surprised, that such an informal Fermi ruling clearly obviated the need for committee discussion or voting.

❖ ❖ ❖

Lincoln Wolfenstein
FERMI INTERACTIONS

For over forty years I have been working in the area of weak interactions. When I begin a course on the subject I write down the interaction Hamiltonian of Fermi in the notation of quantum field theory. This does not seem so impressive to the student, since, in the language of field theory, it says a neutron changes to a proton, and an electron and an antineutrino are emitted, just as postulated by Wolfgang Pauli. To appreciate how significant it was, one needs to look at it in historical perspective. Fermi used to say that once he understood quantum electrodynamics (QED), he knew what to do. Fermi gave lectures on QED at Michigan in the summer of 1931 [1], which remained for many years the clearest presentation of the subject.

The important point is that the beta electron was not there inside the nucleus, but was created, together with the neutrino, by the weak current, just as the electromagnetic current creates a photon in QED. As far as beta decay is concerned, the familiar electron and Pauli's strange particle are equivalent, newly created from the energy of a nuclear transition.

This analogy with QED has proved enormously fruitful and finally, thirty-five years later, has led to what we now call the standard model of weak interactions. In fact, Fermi's original Hamiltonian can still be used to describe the effective interaction at low energies, with one important change: all fields must be replaced with chiral left-handed fields [2]. Fermi's Hamiltonian also made it possible to calculate [3] the interaction cross section of neutrinos, which turned out to be extremely small.

I like to say that this work of Pauli and Fermi marks the beginning of particle physics. The neutrino was the first postulated particle that was not a constituent of matter, and the Fermi weak interaction the first interaction that did not exist in classical physics. Theoretical physicists have been proposing new particles and new interactions ever since, seldom with the success of Pauli and Fermi.

I was at the University of Chicago as an undergraduate in 1942, but I had no idea that Fermi was there. I used to walk down Ellis Avenue past Stagg Field most every day to have lunch at the Ellis Coop; I tell my students that if I had known that the first nuclear reactor was being built there, I would at least have walked on the other side of the street.

When I returned to campus to complete my graduate work in 1946, there were two new professors, Fermi and Edward Teller. No two could have been more different. Fermi's lectures were precise, elegant, and carefully prepared. Teller's emphasized ideas but were often disorganized; on occasion, Frank Yang would emerge from the students to finish a derivation.

A group of students shared a large room on the fourth floor of Eckhart Hall. Mine was the first desk on the right as you came in. Then there were the desks of Geoff Chew and Murph Goldberger, and around the room at least five other students. Fermi worked closely with his students; I remember his coming in and working out the random path of a neutron going through a nucleus, the subject that Murph was to work out for his thesis. On the other hand, students often had trouble getting hold of Teller. I remember once, Fermi and Teller were both in the room; Fermi recognized we were having trouble seeing Teller and said "Edward, you will see this student at 3:30 and this other student at 4:30; all right?" and so, we had appointments with Teller.

On occasion, a group of students would go to lunch with Fermi in the good old Hutchinson Commons. There were two lines: the long line, which allowed for a choice among a number of dishes, and the short line, where the choice was very limited. Fermi always insisted we go in the short line; one didn't waste time waiting in line. Then we would sit somewhere along one of those long tables; I can't remember what we talked about, but it was a real privilege to be sitting there conversing with Enrico Fermi.

After one of Fermi's lectures in the course of quantum mechanics, a student asked, "Professor Fermi, where could I read about what you have lectured about?" Fermi was a little taken aback and finally said, "In any book on quantum mechanics." The student was not satisfied; he wanted a particular book, so Fermi replied, "Name a book on quantum mechanics." The student named "Rojansky." "It must be in Rojansky," Fermi said. Of course it was not in Rojansky; it was in Fermi's notes on quantum mechanics, which are now available in his own handwriting but were not then.

I had a related experience some forty years later. I was giving lectures at the Stanford Linear Accelerator Center (SLAC), discussing the matter effects on neutrino oscillations [4]. I explained that the index of refraction was derived from the optical theorem. At the end a student asked where he could look up the optical theorem, and I replied, "In any book on optics." Luckily, the student didn't ask me to name one, but when I was writing up the lectures, I went to the library and searched several books on optics. However, I could not find the optical theorem. Then I discovered the right reference, where the optical theorem was derived in a very intuitive way by adding the forward-scattered wave to the initial wave to give the transmitted refracted wave. The reference was Fermi's notes on nuclear physics [5]. There he was interested in the index of refraction of neutrons; somewhat surprisingly, it was only thirty years later that I realized the importance of the index of refraction for Fermi's "little neutron," the neutrino.

I left Chicago in the fall of 1948 to go to Carnegie Tech in Pittsburgh, where I have been ever since. However, I came back to Chicago for my thesis defense in the spring of 1949. There I faced my thesis sponsor Teller, Fermi, and Gregor Wentzel. But there was a difficulty: the thesis committee was supposed to have an experimentalist, but Herb Anderson couldn't come at the last minute. "No problem," said Fermi, "I am also the experimentalist, so I will also be the experimentalist on your committee."

I began to make my presentation on polarization effects in nuclear reactions. At some point I explained I would expand the density matrix in terms of a set of tensor operators. Fermi interrupted, "Excuse me, but the experimentalist does not quite understand these tensor operators; will you please explain?" I had a tough experimentalist on my thesis committee, but I did pass.

References

1. E. Fermi, *Reviews of Modern Physics* 4 (1932): 131.

2. R. P. Feynman and M. Gell-Mann, *Physical Review* 109 (1958): 193.

3. H. A. Bethe and R. Peierls, *Nature* 133 (1934): 532.

4. L. Wolfenstein, SLAC Summer School, 1988.

5. E. Fermi, *Nuclear Physics*, ed. J. Orear, A. H. Rosenfeld, and R. A. Schluter (Chicago: University of Chicago Press, 1950), p. 201.

❖ ❖ ❖

Chen Ning Yang
REMINISCENCES OF ENRICO FERMI

They that have power to hurt and will do none,
That do not do the thing they most do show,
Who, moving others, are themselves as stone,
Unmoved, cold, and to temptation slow;
They rightly do inherit heaven's graces,
And husband nature's riches from expense;
They are the lords and owners of their faces
                                    William Shakespeare, sonnet 94

Enrico Fermi was, of all the great physicists of the twentieth century, among the most respected and admired. He was respected and admired because of his contributions to both theoretical and experimental physics, because of his leadership in discovering for mankind a powerful new source of energy, and above all, because of his personal character: He was always reliable and trustworthy. He had both his feet on the ground all the time. He had great strength, but never threw his weight around. He did not play to the gallery. He did not practice one-upmanship. He exemplified, I always believe, the perfect Confucian gentleman.

Fermi's earliest interests in physics seem to be in general relativity. Starting from around 1923 he began to think deeply about the "Gibbs paradox" and the "absolute entropy constant" in statistical mechanics. Then, as Emilio Segrè later wrote, "As soon as he read Pauli's article on the exclusion principle, he realized that he now possessed all the elements for a theory of the ideal gas which would satisfy the Nernst principle at the absolute zero, give the correct Sackur-Tetrode formula for the absolute entropy in the limit for low density and high temperature, and be free of the various arbitrary assumptions that it had been necessary to introduce in statistical mechanics in order to derive a correct entropy value" [1].

This research led to his first monumental work, and to the "Fermi distribution," "Fermi sphere," "Fermi liquid," "Fermions," and so on.

It was characteristic of Fermi's style in research that he should follow this abstract contribution with an application to the heavy atom, leading to what is now known as the Thomas-Fermi method. The differential equation involved in this method was solved by Fermi "numerically with a small and primitive hand calculator. This numerical work took him probably a week. E. Majorana, who was a lightning-fast calculator and a very skeptical man, decided to check the numerical work. He did this by transforming the equation into a Riccati equation and solving the latter numerically. The result agreed exactly with the one obtained by Fermi" [2].

Fermi's love of the use of computers, small and large, which we graduate students at Chicago observed and admired, began evidently early in his career and lasted throughout his later life.

Fermi's next major contribution was in quantum electrodynamics, where he succeeded in eliminating the longitudinal field to arrive at the Coulomb interaction. Fermi was very proud of this work, as his students at the University of Chicago in the years 1946–54 knew. (But it seems today that few theorists under the age of sixty-five know about this contribution of Fermi's.) It again was characteristic of Fermi's style that in this work he saw through complicated formalisms to arrive at the basics, in this case a collection of harmonic oscillators, and to proceed to solve a simple Schrödinger-like equation. The work was first presented in April 1929 in

Paris and later at the famous summer school at Ann Arbor in the summer of 1930. G. Uhlenbeck told me in the late 1950s that before this work of Fermi, nobody had really understood the quantum theory of radiation, and that this work had established Fermi as among the few top field theorists in the world.

I shall skip describing his beautiful contribution in 1930 to the theory of hyperfine structure and come to the theory of β-decay. According to Segrè, Fermi had considered, throughout his life, that this theory was his most important contribution to theoretical physics. I had read Segrè's remarks in this regard, but was puzzled. One day in the 1970s, I had the following conversation with Eugene Wigner in the cafeteria of Rockefeller University:

Y: What do you think was Fermi's most important contribution to theoretical physics?

W: β-decay theory.

Y: How could that be? It is being replaced by more fundamental ideas. Of course it was a very important contribution which had sustained the whole field for some forty years: Fermi had characteristically swept what was unknowable at that time under the rug, and focused on what can be calculated. It was beautiful and agreed with experiment. But it was not permanent. In contrast, the Fermi distribution is permanent.

W: No, no, you do not understand the impact it produced at the time. Von Neumann and I had been thinking about β-decay for a long time, as did everybody else. We simply did not know how to create an electron in a nucleus.

Y: Fermi knew how to do that by using a second quantized ψ?

W: Yes.

Y: But it was you and Jordan who had first invented the second quantized ψ.

W: Yes, yes. But we never dreamed that it could be used in real physics.

I shall not go into Fermi's later contributions. Nor into his relations with students, which I have written about before.[3] I shall only add a couple of stories about Fermi.

One of Fermi's assistants at Los Alamos during the war was Joan Hinton, who became a graduate student at the University of Chicago after the war. When I began working in late 1946 for Sam Allison, she was a fellow graduate student in the same laboratory. In the spring of 1948 she went to China and married her boy friend Sid Engst and settled down in China permanently to do agricultural work. (Hers was a very interesting story that should be written down. I hope she will do it soon.) In the summer of 1971, during my first visit to the New China, half a year before Nixon, I accidentally met her in a hostel in Da-zhai, then a model agricultural commune in the county of Xi-Yang. Surprised and delighted, we reminisced about the Chicago days: how I was awkward in the laboratory; how I almost accidentally electrocuted her; how I had taught her a few sentences of Chinese; how I had borrowed a car and had driven her to the LaSalle Street Station to embark on her long trip to China; and so on. She asked me whether I remember the farewell party that the Fermis had given her before she left. I did. Did I remember the camera that they had given her that evening? No, I did not. After a pause, she said she had felt, a few days before that farewell party, that she should tell Fermi about her plan to go to the communist-controlled area of China. So she did. And what did Fermi say? "He did not object. For that I am eternally grateful." I considered this such an important statement[4] that after coming back to Stony Brook, I called Mrs. Fermi in Chicago and reported to her my whole encounter with Joan in Da-zhai. A few years later, Joan visited Chicago herself and had the opportunity to visit with Mrs. Fermi and her daughter, Nella Fermi.

I shall end this article by quoting from my *Selected Papers* (1983) [5]: "Fermi was deeply respected by all, as a physicist and as a person. The quality about him that commands respect is, I believe, solidity. There was nothing about him that did not radiate this fundamental strength of character. One day in the early 1950's, J. R. Oppenheimer, who was the Chairman of the important General Advisory Committee (GAC) of the Atomic Energy Commission (AEC), told me that he had tried to persuade Fermi to stay on the GAC when Fermi's term was up. Fermi was reluctant. He pressed, and finally Fermi said, 'You know, I don't always trust my opinions about these political matters.'"

Notes and References

1. E. Segrè, in *Collected Papers of Enrico Fermi* (Chicago: University of Chicago Press, 1962), p. 178.

2. Ibid., p. 277.

3. Ibid., p. 673.

4. Joan Hinton went to China in the spring of 1948, before the Chinese communists' victory over Chiang Kai-Shek, and two years before the Korean War. If she had planned to go to China after the beginning of the Korean War, I am sure the U.S. government would not have allowed her to go.

5. C. N. Yang, *Selected Papers, 1945–1980* (San Francisco, W. H. Freeman, 1983), p. 48.

❖ ❖ ❖

Gaurang Yodh

THIS ACCOUNT IS NOT ACCORDING TO THE *MAHABHARATA*!

How I Chose to Come to Chicago

As an undergraduate in Bombay, I was interested in pursuing graduate study in nuclear physics at the best possible place. The Tata Institute of Fundamental Research had just opened in 1945, and I had the opportunity to listen to lectures on geomagnetic effects on cosmic rays by Manuel Vallarta. I got a chance to talk with a professor there, P. S. Gill, who had studied in the States and had done experimental work in cosmic rays and asked for his suggestions as to where to apply. He directed me to write to a Professor James Benade in Lahore, who had worked with Arthur Compton and said he would be a valuable resource. I wrote to him for advice and received a reply from him which said in essence, "Go to Chicago, it is the best place for nuclear studies, as Enrico Fermi is there." I applied to Chicago and was admitted while still doing my B.S. in Bombay.

First Encounter with Fermi

At Chicago, one of my first assistantships was with Professor Robert Mulliken, to work on an infrared Perkin Elmer spectrometer in the basement of Eckart Hall. Although I did not stay with it for more than a year, it gave me a first feel of experimental work with total independence! My first interaction with Fermi was in 1949, when I was at a party given by Robert Mulliken in his apartment on Blackstone. I was requested to play sitar for the group, which included Fermi, Wentzel, and other distinguished faculty. After my performance, Fermi came up to me and asked "Gaurang, may I ask you a question?" I was pleased and said yes. He continued: "I am not musically inclined but am interested in your opinion whether you believe instruments improve with age, and if so, why?" It was wonderful to be asked a precise question, and I answered: "Yes they do for a period of time, but then they have to be worked on."

Origin of Cosmic Rays

Fermi had a suspicion that magnetic fields could accelerate cosmic rays. In 1948 Hannes Alfvén visited Chicago to discuss with Edward Teller the question of the solar origin of cosmic rays. Fermi learned from Alfvén about the probable existence of extended magnetic fields in the galaxy. He figured that these fields would necessarily be dragged around by the ionized interstellar medium and would result in randomly moving magnetic clouds. Fermi realized that here was an excellent way of acceleration of cosmic rays throughout the galaxy. This led to his seminal paper in the *Physical Review*, "On the Origin of the Cosmic Rays," where cosmic rays were accelerated by stochastic processes. I still remember vividly the lucid colloquium he gave on first- and second-order acceleration of charged particles colliding with these magnetic clouds. [Fermi's original notes on cosmic ray acceleration are presented in chapter 5.—Ed.]

Elementary Particles circa 1950

Fermi gave a set of lectures in Rome about elementary particles. At this time the particles which seemed to be elementary were the electron, proton, neutron, pion, muon, neutrino, and photon. In order to prepare for the possibility of doing experiments with pions Fermi developed a framework to describe the four forces of nature then known and, in particular, electric, strong, and weak forces, which made it possible to estimate rates and cross sections of various processes with elementary particles. These were given as the Silliman Lectures on nuclear and particle physics at Yale [1]. He estimated in the true Fermi approach the strengths of weak interaction from properties of muon decay and capture and pion decay. He estimated the strength of strong interactions from the then measured pion production and pion capture reactions.

Foray into Extremely High Energies

One of the first jobs I was able to get at Chicago in particle physics (at 75¢/hour) was with Jere Lord, who was a student of Marcel Schein. I had to scan emulsions for determining the multiplicity of 20-GeV cosmic ray interactions. This group flew emulsion stacks with payloads attached to rubber balloons from Stagg Field! An extremely collimated event was seen in cosmic rays at Chicago by the group of Schein [2]. The story is that Jere Lord, who was a postdoc with Schein and who had been his student, and Joseph Fainberg detected, in emulsion, a very unusual event, now called the Schein star, which had a release of 200 GeV of energy and a symmetric particle production in the center of mass with highly collimated beams forward and aft, as shown in figure 8.6.

Jere Lord recognized the unusual nature of the event and, after measurement of the event, showed the event to Fermi before he showed it to Schein, and pointed out its unusual characteristics. Fermi at that time had developed the statistical model for multiparticle production at very high energies and immediately went on to incorporate angular momentum into

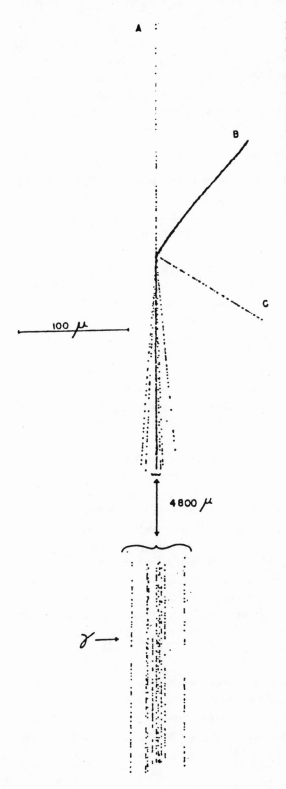

Figure 8.6 Microprojection drawing of the Schein star produced by a proton with 3×10^{13} eV. The particles are produced in two cones in the center of mass of the collision. The backward cone corresponds to the large-angle tracks coming from the interaction vertex. The forward cone is resolved only by following the tracks 4800 microns downstream. (Figure from J. J. Lord, J. Fainberg, and M. Schein, *Physical Review* 80 [1950]: 970. Copyright 1950 by the American Physical Society.)

the original picture to explain the sharply peaked Schein star. Although the statistical model for high-energy multiparticle production is no longer the favored picture, the thermodynamic character of the model is very general and is invoked even today in the search for a phase transition in the quark-gluon system.

Research on the Synchrocyclotron

When I passed the basic examination, I first went to Professor Gregor Wentzel to discuss working in theory. He asked me whether I had any interest in experiment? He went on to say that there was no satisfactory theory of strong interactions and theoreticians needed good experimental guidance.

I liked experiments, so I went to Fermi. Fermi said this was a very good time to join work on the synchrocyclotron and study pion physics. So, I joined his group—Herbert Anderson, Darragh Nagle, Arné Lundby, and Ronald Martin. My earliest jobs were to make cables and put on the connectors, make scintillation counters, and help lay out the pion channels through the shielding wall. Fermi gave a seminar on the technique of floating wires to determine pion trajectories.

In a magnetic field H, a wire under a tension T and carrying a current I will follow exactly the orbit of a particle with magnetic rigidity $H \times \rho = T/I$. [Fermi also used the Maniac to calculate these orbits; see chapter 5.—Ed.] The floating wire measurements required me to crawl between the cyclotron pole faces and record the angle at which the floating wire exited the vacuum chamber when it was attached to the target on Fermi's famous trolley.

When the cyclotron first went into operation, there was no external beam, and therefore it was important to determine the radial structure of the internal beam. This was done by activation methods. Aluminum foils were exposed to the internal beam at the end of a probe into the vacuum chamber at different radial distances and their activation measured. My job was to take the exposed foil and run up to the room where Fermi was sitting next to a Geiger counter setup, with his slide rule in hand. Fermi determined its radioactivity while we went down and exposed another foil at another radial distance. This is the only time I observed that Fermi stayed up working up to midnight; otherwise he promptly finished his work at the institute at 6 p.m.

Once the pion beams were available, Fermi and his team set up the scattering experiments. First we measured total cross sections in various elements including heavy water using a poor geometry arrangement. Then we started on the famous experiments on pion-hydrogen scattering. During this time, before we started the charge exchange measurements, Fermi came to know that Richard Dalitz had calculated the internal conversion of pi-zero gamma rays. He invited us into his office to show us a derivation, ab initio, of the reported calculation of this Dalitz pair rate! During this time I worked on development and assembling of several different liquid scintillator counters, using obnoxious and probably cancerous chemicals.

An International Conference on Nuclear Physics and Fundamental Particles was held in Chicago in September of 1951 in the Oriental Institute. Fermi gave the inaugural talk on the status of elementary particles. The first page of Fermi's remarks is shown in figure 8.7. These notes were typed up by some of Fermi's graduate students

My Thesis Research

At this time, students were encouraged to do their thesis research entirely on their own. My first attempt at a thesis was to study μ-mesic x-rays with a 1-inch sodium iodide crystal and a large ten-channel analyzer—a gift of Dick Garwin. I was just about ready to take data when Fermi returned from Columbia and said, "Gaurang, your thesis has just been done at Columbia by Fitch and Rainwater!" Such was the keen competition between two great institutions. I was very disappointed at being scooped, and searched for another topic. I decided to study charged pion production in neutron-proton collision. With the guidance of Roger Hildebrand, we developed a lethal 400-MeV neutron beam. After a couple of false starts, with help of Arthur Rosenfeld I decided upon using emulsions for detecting the pions from this reaction.

After completing the work and writing up the results in my thesis, I gave a copy to Fermi when he was about to leave for Varenna in 1954. He said to me that it looked good but that he would check it. I did not know what he thought of it until early 1955, when his Varenna lectures on pion physics came out, and he referred to my work as the "Yodh technique" and presented the results as shown in figure 8.8 [3].

Fermi as a Teacher

He was a dedicated and excellent teacher at all levels. He taught physics to first-year students in the college, and gave a full course of lectures in physics from mathematical physics, thermodynamics and statistical physics, quantum mechanics, to nuclear physics, and particle physics for graduate students. These lectures were superb in their clarity, breadth, and depth. We all attended them, even if we had taken the course before!

Whenever Fermi learned something new he gave a set of lectures to

SESSION I
PRESENT STATUS OF KNOWLEDGE CONCERNING FUNDAMENTAL PARTICLES

Speaker   Enrico Fermi
Topic   "FUNDAMENTAL PARTICLES"

E. Fermi presented the following list of 21 elementary particles.

e electron	μ^- negative muon
e^+ positron	μ^+ positive muon
P proton	G graviton
$\bar{P}$ antiproton	V^+ positive V-particle
N neutron	V^- negative V-particle
$\bar{N}$ antineutron	V^0 neutral V-particle
γ photon	τ^+ positive τ-meson
π^+ positive pion	τ^- negative τ-meson
π^- negative pion	κ^+ positive κ-meson
π^0 neutral pion	κ^- negative κ-meson
	ν neutrino

Fermi expressed a belief in the existence of antinucleons. He defined the V, κ, and τ particles during the discussion period (see page 4).

Philosofically, at least some of these 21 particles must be far from elementary. The requirement for a particle to be elementary is that it be structureless. Probably some of these 21 particles are not structureless objects. They may even have some geometrical structure, if geometry has any meaning in such a small domain.

Fundamental particles are distinguished most easily by charge, mass, and spin. The spin of the positive pion has recently been determined by cyclotron results. This utilizes the principle of detailed balancing. Thus the result does not depend on any knowledge of the types of interactions involved. The reaction utilized is

I.1 $\pi^+ + D \rightleftarrows P + P$

I.2 $(2I_\pi + 1)(2I_D + 1)\, \sigma\, p^2 = (2I_p + 1)(2I_p + 1)\, \sigma\, p_2^2 \times \tfrac{1}{2}$

$3(2I_\pi + 1)\sigma\, p^2 = 2\, \sigma\, p_2^2$

The factor ½ on the right hand side of eq. I.2 is because the protons are identical particles (an interchange of the two protons does not give a new state). The experimental values give $2I + 1 \approx 1$ so $I = 0$. Since the π^+ and π^- are considered similar, both should have spin zero.

Recent results have shown that the pion is pseudoscalar. The parity of a particle is one of the most fundamental distinctions. Once we have decided the parity, we limit roughly by ½ the number of possible interactions. It should be noted that whether the pion is a true or a pseudoscalar has no meaning for the pion by itself. It has meaning only in relation to interactions with other particles; for example, in $P \rightleftarrows N + \pi^+$.

1

Figure 8.7 First page of notes of Fermi's lecture on fundamental particles, at the International Conference on Nuclear Physics and the Physics of Fundamental Particles, held in Chicago in 1951. Notes taken by Fermi's students are reproduced in Enrico Fermi, *Collected Papers*, volume 2 (Chicago: University of Chicago Press, 1965).

students and colleagues. One such time was when he returned after spending some months working on the Los Alamos Maniac with N. Metropolis. He gave a series of talks on how to use the Maniac, which was our first introduction to computing.

Fermi Vignettes

Fermi was simple and direct. He loved to watch thunderstorms, and so did I. Chicago in the summer has many of them. Once we were watching thunderstorms outside the institutes and I asked him, "Why did you develop the atom bomb?" He answered simply, "We had to be better than Hitler and defeat him."

Once, while he was watching me compute in the cyclotron control room, he said, "Gaurang, you know Herb computes ten times faster than you, and I can compute ten times faster than Herb!!" He said it in such a simple and straightforward manner that I accepted it as fact rather than feeling depressed.

When Fermi had to go out of town, he usually asked C. N. Yang or Marvin Goldberger to give his lecture. He gave them a carefully worked-out set of notes for the lecture indicating what he intended to cover and how. In his lectures, Fermi never used the sentence "It is well known . . ." or "It is obvious . . ." The lucidity of his lectures is legendary.

During our experiments on pion scattering, Fermi was a member of a book club and was reading the Indian epic *Mahabharata*. Once, while I was adjusting a counter's position or orientation, he came up to me and said, "That is not the way it is done in the *Mahabharata*."

Fermi treated his students as his extended family. He played games at parties they held at their house and, of course, liked to win. He liked the

---

LECTURES ON PIONS AND NUCLEONS   71

ing. These are due to i) YODH (unpublished), ii) WRIGHT and SCHLU-
TER ([13]).

i) The Yodh technique has been used in studying the two np reactions in which positive and negative pions emerge. A liquid hydrogen target is used and photographic plates are suitably positioned to record the pion tracks, as shown in Fig. 16. In this manner the total cross-section and angular distribution can be studied, for the reactions

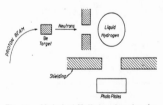

$$np \left\{ \begin{array}{l} pp - \\ nn + \end{array} \right.$$

Fig. 16. – Method of Yodh for observing the pions produced in n-p collisions.

ii) WRIGHT and SCHLUTER have used, instead, a hydrogen filled diffusion cloud chamber operating at a pressure of 30 atmospheres in conjunction with a magnetic field. With the neutron beam, one can use a high flux of primary neutrons since they are not detected in the chamber unless they interact. Most of the events

Figure 8.8 A page from Fermi's lectures on pion physics at the 1954 Varenna meeting, the summer before his death. (Reproduced with permission of *Nuovo Cimento*.)

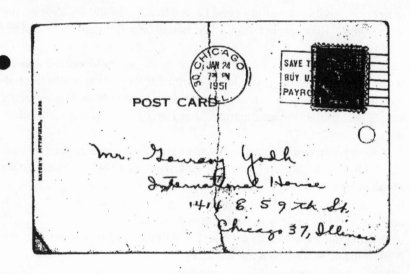

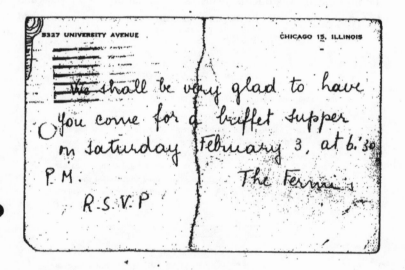

Figure 8.9 An invitation to supper, written by Laura Fermi.

annual physics Christmas parties, and enjoyed the skits we performed and the songs we composed and sang. An invitation to supper at the Fermis', written on a postcard, is shown in figure 8.9. Here also are a couple of stanzas of the "Nobel Prize Song," presented at a 1954 Christmas party. It was composed by Arthur Rosenfeld and set to the tune of "Waltzing Matilda."

Nobel Prize Song
Art and Bob and Jay scan lots of plates and photographs
Looking for pairs and mu's and pi's.
"Thank God," says Fermi, "I need not depend on them.
I already have my Nobel Prize."
Nobel Prize, Nobel Prize,
I already have my Nobel Prize.
"Thank God," says Fermi, "I had better men than they.
They've already got me my Nobel Prize."

Megavolts give you jolts, circuitry requires ability,
Theory is just a pack of lies.
Solid state ain't so great, gamma rays may cause sterility
And so to hell with the Nobel Prize.
I am tired, I am sleepy
I don't want any Nobel Prize.
Let's get away from this radioactivity,
And so to hell with the Nobel Prize (in physics).

Notes and References

1. E. Fermi, *Elementary Particles* (New Haven, Conn.: Yale University Press, 1951)

2. J. Lord, J. Fainberg, and M. Schein, *Physical Review* 80 (1950): 970.

3. E. Fermi, "Lectures on Pions and Nucleons," ed. B. T. Feld, *Nuovo Cimento*, Suppl. 2 (1955): 17–95.

Chapter Nine

What Can We Learn with
High Energy Accelerators?

❖ ❖ ❖

James W. Cronin

FERMI'S LOOK INTO HIS CRYSTAL BALL

As soon as Fermi assumed his professorship at the newly
created Institute for Nuclear Studies, he laid out his vision
for future research in a letter to the dean of Physical Sciences, Walter Bartky (chapter 5). The main thrust of his
research plan was a move into particle physics employing
the highest energies possible. The motivation was to
artificially produce mesons. In 1945, the pion had not
been discovered, and it had yet to be shown that the observed meson (now the muon) was weakly interacting.
The recognition of a weakly interacting muon, and the
discovery of the strongly interacting pion, came in rapid
succession in 1947. Fermi had requested a 100-MeV betatron, the highest-energy accelerator available in 1945.
The principle of phase stability was soon discovered,
which allowed the construction of a 450-MeV cyclotron
at the institute and the classic pion scattering experiments
of Fermi and his collaborators.

Fermi's vision extended well beyond the study of pions. By the early fifties, discoveries were made in cosmic
rays of heavier mesons and heavier baryons. The field of
particle physics was about to embark on a golden age. The

key to the golden age was the development of new accelerators of higher energy. In 1953, the 3-GeV Cosmotron started to operate at Brookhaven National Laboratory, and the 6 GeV-Bevatron was under construction at Berkeley. (Note that, at Fermi's time, the energy was expressed as "BeV." Only when the European accelerators began to come was BeV replaced by GeV). Fermi chose the title "What Can We Learn with High Energy Accelerators?" as the subject of his address as retiring president of the American Physical Society. The speech was delivered on January 29, 1954. There is no known recording of this speech; however, in the University of Chicago archives there was an outline of the speech written in pencil by Fermi, followed by an outline typed by Fermi. Also, the six slides he showed were in the archives. These outlines are reproduced in figures 9.1 and 9.2.

Fermi points out that the only way to access this new physics is through the development of higher- and higher-energy accelerators. In his first slide (fig. 9.3), he plots the growth with time of the energy of particle accelerators beginning with the 1-MeV Cockcroft-Walton accelerators of 1930. The final solid point is the 3-GeV Cosmotron, which began operation in 1953. The open circles represent the Berkeley Bevatron and the Brookhaven AGS, which was in the planning stage at the time. Fermi extrapolated his curve to the "ultimate accelerator," one that circled the globe at the equator. We refer to this accelerator as Fermi's "Globatron." Following Fermi's extrapolation, the Globatron would have arrived in 1994 with an energy of 5×10^9 MeV, or 5×10^{15} eV. A dramatic view of this accelerator is shown in Fermi's second slide (fig. 9.4). In his first slide (fig. 9.3), Fermi also plots the evolution of cost of the accelerators. The cost of the ultimate accelerator was 170 billion 1954 dollars!

As to the increase in energy with time, Fermi's curve was remarkably correct. In the nearly fifty years that have passed we can plot the actual progress of accelerators and compare to Fermi's curve. It would have seemed impossible to remain on Fermi's curve to the end of the twentieth century. Nevertheless, it turned out that Fermi's extrapolation was remarkably close to the actual progression. This fact was due to the development of the technology of colliding beams and the development of superconducting magnets. In figure 9.5, we plot the actual growth of accelerators. We plot only hadron machines and convert the energy of the center of mass of the colliding beam machines into equivalent energy of a single beam on a fixed target. The 400-GeV accelerator at Fermilab falls below the curve, while the CERN 30-GeV × 30-GeV Intersecting Storage Ring (ISR) maintains the progression. One can also note that the ill-fated Superconducting Super Collider (SSC) would have been a factor of 10 ahead of its time, while the

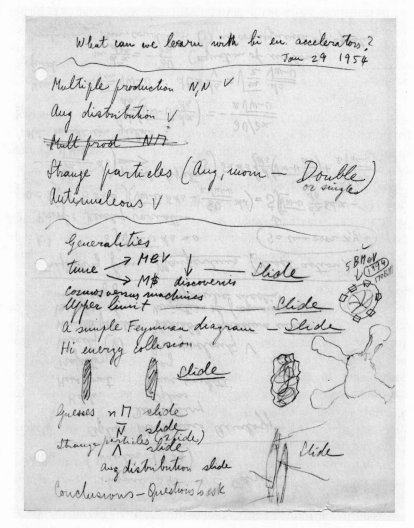

Figure 9.1 Fermi's penciled notes "What can we learn with high energy accelerators?" (Reproduced with permission of Special Collections, University of Chicago Libraries.)

CERN Large Hadron Collider (LHC) is a bit behind. We do not know whether Fermi had a remarkable intuition or was "pulling our leg," but we cannot accuse him of not thinking big.

Not included in figure 9.5 are the electron-positron colliders, which are a bit difficult to add to the graph without a more technical discussion, which is beyond the scope of this essay.

In figure 9.6, we have plotted the costs of the accelerators as a function

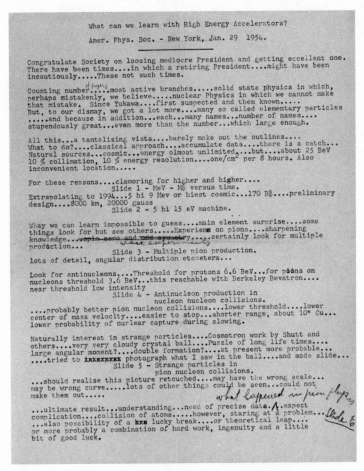

Figure 9.2 Fermi's typed notes "What can we learn with high energy accelerators?" (Reproduced with permission of Special Collections, University of Chicago Libraries.)

of time [1]. The costs fall dramatically below Fermi's projection. Again this is due to the development of colliding beams and superconducting magnets. The costs in figure 9.6 should be taken as a trend and not with the precision of an accountant. Inflation is not considered, nor is the fact that all the accelerators built since 1980 have been attached to existing infrastructure.

Fermi was correct in the prediction that the way to discovery required higher energies. He says in the outline, "what we can learn impossible to guess"; "some things to look for"; "certainly look for multiple production."

Fermi was particularly interested in the early results of multiple meson production, which were observed at the Brookhaven Cosmotron. He had

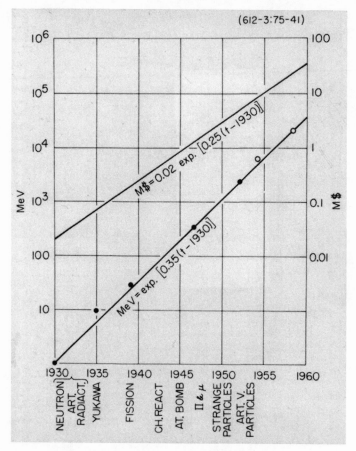

Figure 9.3 Fermi's slide 1, cost and energy of accelerators as a function of time. (Reproduced with permission of Special Collections, University of Chicago Libraries.)

developed a statistical model for the multiple production of particles in 1950 [2]. While later authors took some pedantic glee in showing that Fermi's predictions were not exactly correct, Fermi never intended the calculations to do more than give an order of magnitude of the particle production. The next three slides represent Fermi's anticipation of what would be found with the accelerators. Fermi presents his expectation for multiple meson production in nucleon-nucleon collisions as a function of energy in his third slide (fig. 9.7). In figure 9.8 we have computed the mean multiplicity from Fermi's third slide and compared it with the mean multiplicity curve actually measured [3]. The agreement is remarkable.

Following the outline, Fermi then spoke of the production of

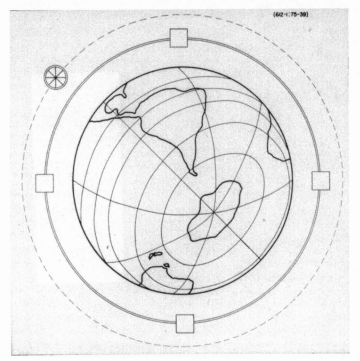

(6l2-1:75-39)

Figure 9.4 Fermi's slide 2, the "Globatron." (Reproduced with permission of Special Collections, University of Chicago Libraries.)

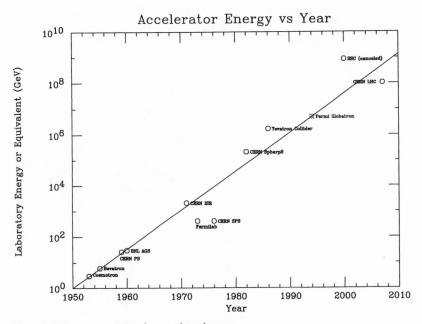

Figure 9.5 Energy versus time for actual accelerators.

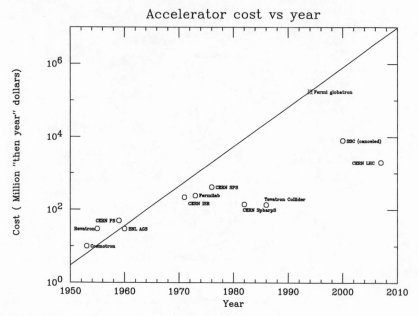

Figure 9.6 Approximate cost of actual accelerators versus time.

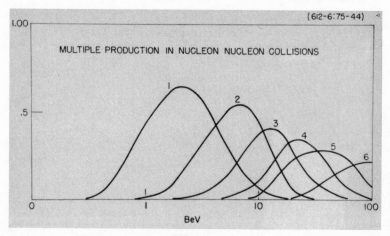

Figure 9.7 Fermi's slide 3, multiple meson production. (Reproduced with permission of Special Collections, University of Chicago Libraries.)

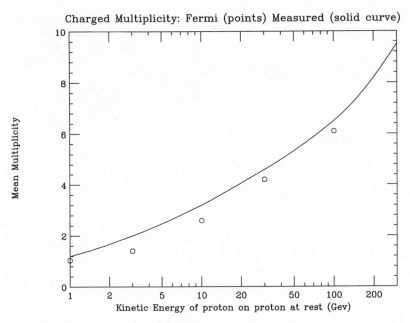

Figure 9.8 Measured charged multiplicity versus energy.

antinucleons. This was at a time before the operation of the Berkeley Bevatron and the discovery of the antiproton. In Fermi's fourth slide (fig. 9.9) he presents his calculation of antinucleon production in proton-proton collisions as a function of energy. In figure 9.10, we compare Fermi's predictions with the measured antiproton production data. A direct measurement of the total production cross section for antiprotons has never been made. The total production must be inferred from a model normalized by the few measurements that exist for production at specific energies, angles, and incident energy. The points less than 10 GeV are near threshold on a nuclear target, so the nuclear motion, "Fermi momentum," must also be included to extract the production fraction [4]. The data at higher energy derive from an empirical fit to antiproton production for the purpose of building an antiproton source for the Tevatron collider [5]. Fermi's anticipation of the antiproton production was remarkably accurate.

Fermi then passes to a discussion of the production of strange particles. Strange particles were for the first time being produced at the Cosmotron accelerator. He refers to looking into his crystal ball: "very very cloudy." Out of the ball comes Fermi's fifth slide (fig. 9.11), a prediction of the strange particle fraction to be found in pion-nucleon collisions. His notes read, "may have wrong scale"; "may be wrong curve." In figure 9.12, we plot the

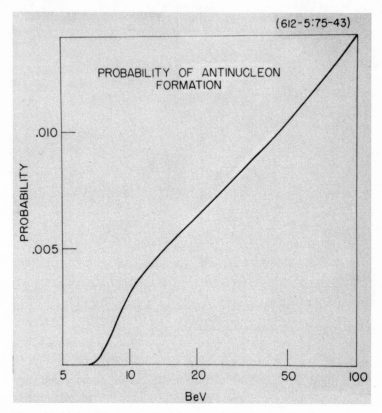

Figure 9.9 Fermi's slide 4, antinucleon formation. (Reproduced with permission of Special Collections, University of Chicago Libraries.)

measured production and Fermi's view from the crystal ball. His view has been remarkably clear!

Fermi's last slide (fig. 9.13) shows a Feynman graph for "the 22nd approximation" for the production of three nucleons, one antinucleon, and two pions in a nucleon-nucleon collision. This slide, added as an afterthought to his typed outline, expresses his exasperation with the strong interaction theories of the time and the hopelessness of doing any detailed calculations. Thus, he returned to the statistical physics that marked his early career in Italy to anticipate many phenomena in high-energy physics.

In his retiring lecture Fermi clearly anticipated the excitement of particle physics that lay ahead. But even Fermi would have been surprised at the outcome of the following years. No one could have anticipated all the surprises and understanding in the next thirty years. It is such a pity that

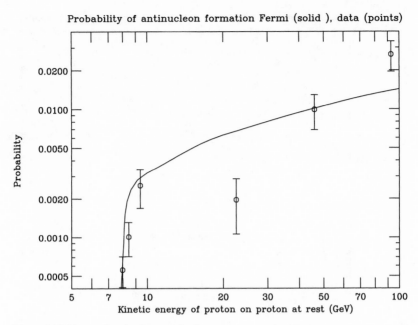

Figure 9.10 Measured antinucleon production.

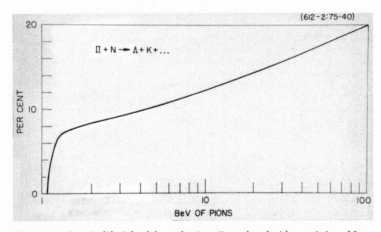

Figure 9.11 Fermi's slide 5, lambda production. (Reproduced with permission of Special Collections, University of Chicago Libraries.)

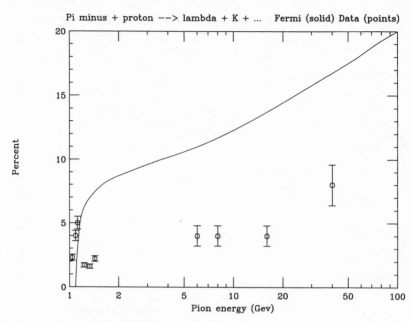

Figure 9.12 Measured lambda production.

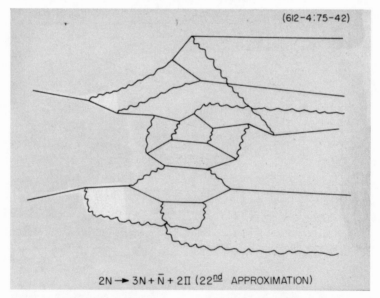

Figure 9.13 Fermi's slide 6, "22nd approximation." (Reproduced with permission of Special Collections, University of Chicago Libraries.)

Fermi did not live through this golden age. One wonders what his contributions might have been.

References and Notes

1. The cost figures were found in an unpublished "Catalog of High Energy Accelerators" in the CERN Library. I want to thank Adrienne Kolb of the Fermi National Accelerator Laboratory for her assistance in locating this material.

2. E. Fermi, *Progress of Theoretical Physics* (Japan) 5 (1950): 570.

3. Most of the data used for the comparisons with Fermi come from databases maintained by the COMPAS Group (Russia) at the Web site http://wwwp-pds.ihep.su:8001/ppds.html, and the HEPDATA database maintained by the University of Durham/RAL at the Web site http://durpgd.dur.ac.uk/HEPDATA/REAC.

4. The threshold data come from measurements published in T. Elioff et al., *Physical Review* 128 (1962): 869. The total production probability is extracted using the model found in O. Chamberlain et al., *Nuovo Cimento* 3 (1956): 462.

5. C. Hojvat and A. Van Ginneken, *Nuclear Instruments and Methods* 206 (1983): 67.

Further Reading

1. Bethe, Hans, Silvan Schweber, Valentine Telegdi, and Roy Glauber. Articles on Fermi in *Physics Today,* vol. 55 (June 2002). These articles were written to celebrate Fermi's hundredth anniversary.

2. Fermi, Enrico. *Enrico Fermi: Collected Papers.* Edited by Emilio Segrè. 2 vols. University of Chicago Press, 1962, 1965. These volumes are of great interest for the physicist. For most of the papers there is a commentary by a contemporary. It also contains many unpublished papers and documents.

3. Fermi, Laura. *Atoms in the Family.* University of Chicago Press, 1954; reprinted by the American Institute of Physics, 1987. A delightful view of Fermi from a unique perspective.

4. Latil, Pierre de. *Enrico Fermi: The Man and His Theories.* Souvenir Press, 1964. A brief and popular biography of Fermi, translated from French.

5. Maltese, Giulio. *Enrico Fermi in America: Una biografia scientifica, 1938–1954.* Zanichelli, 2003. A new scientific biography of Fermi in Italian.

6. Orear, Jay, et al. *Enrico Fermi: The Master Scientist.* Internet–First University Press, 2004. A personal view of Fermi with contributions from distinguished colleagues.

7. Sachs, Robert G., ed. *The Nuclear Chain Reaction—Forty Years Later: Proceedings of a University of Chicago Commemorative Symposium.* University of Chicago, 1984.

8. Segrè, Emilio. *Enrico Fermi, Physicist.* University of Chicago Press, 1970. This is the most authoritative biography of Fermi presently available.

Contributors

Harold Agnew

Harold Agnew was born in Denver, Colorado, in 1921. He received a B.A. in chemistry from the University of Denver in 1942. He joined Fermi's research group at Chicago in 1942. He was sent to Columbia and then moved with Fermi back to Chicago and participated in the construction of the pile under the west stands of Stagg Field. He was a witness at the initiation of the first controlled nuclear chain reaction on December 2, 1942. Following this event he moved to Los Alamos in 1943. On August 6, 1945, he flew with the 509th Composite Group to Hiroshima with Luis Alvarez and measured, from the air, over the target, the yield of the first use of the atomic bomb. In 1946 he returned to Chicago to complete his graduate studies and received a Ph.D. in 1949 under Fermi's direction. Following his stay at Chicago he returned to Los Alamos in the Physics Division and eventually became the Weapons Division leader (1964–70). In 1970 he became director of the Los Alamos Scientific Laboratory. In 1979 he retired and became president of General Atomics and retired in 1983. He was scientific adviser to SACEUR at NATO (1961–64), a member of the President's Science Advisory Committee (1965–73), and a White House science councilor (1982–89). He was chairman of the General Advisory Committee of the Arms Control and Disarmament Agency (1974–78). He also had a political career, being a New Mexico state senator from 1955 to 1961, when he resigned to join NATO. He has received many recognitions for his service, including the E. O. Lawrence Award in 1966 and the Enrico

Fermi Award of the Department of Energy in 1978. He is presently adjunct professor at the University of California, San Diego.

Nina Byers
Nina Byers was born in Los Angeles, California, in 1930. She received a B.A. from the University of California, Berkeley, in 1950. She entered graduate school at the University of Chicago in 1951 and received her Ph.D. under the direction of Gregor Wentzel in 1956. She was a postdoctoral fellow at the University of Birmingham, United Kingdom, from 1956 to 1958, working with Rudolph Peierls. Following a three-year faculty appointment at Stanford University she joined the faculty of the University of California at Los Angeles. During the years 1967–76 she was also a fellow of Somerville College, Oxford University. Her research covers diverse topics in the theory of particle physics and the theory of superconductivity. She has held research appointments at the European Organization for Nuclear Research and the Fermi National Accelerator Laboratory. More recently her interests have turned to the history of physics with emphasis on the contribution of twentieth-century women to physics. She served as a councillor of the American Physical Society (APS) from 1977 to 1981 and has served as a member of the Panel on Public Affairs and the Forum on History of Physics of the APS. She is a member of the Federation of American Scientists.

Owen Chamberlain
Owen Chamberlain was born in San Francisco, California, in 1920. He received a B.A. in physics from Dartmouth College in 1941. Following his undergraduate degree he joined the Manhattan Project as a civilian, first at Berkeley and then at Los Alamos, where he encountered Fermi. During this time he with Emilio Segrè and George Farwell discovered that plutonium-240 produced in the Hanford pile along with the explosive agent plutonium-239 had a high rate of spontaneous fission which demanded an implosion mechanism to trigger the plutonium bomb. After the war he entered the graduate school at Chicago and completed a Ph.D. under the direction of Fermi in 1949. He joined the physics department at Berkeley and began a long career of research using the Berkeley cyclotron and then the Bevatron. He had a lifelong interest in polarization and made some of the first studies of the polarization of protons produced in elastic scattering with Segrè and collaborators. He also completed a series of triple scattering experiments.

In 1956, when the Bevatron came into operation, Chamberlain and colleagues observed the production of antiprotons. For this work Chamberlain and Segrè were awarded the Nobel Prize in 1959. Following the antiproton work, Chamberlain developed practical polarized proton targets with Carson Jeffries and carried out many experiments with them.

Geoffrey Chew
Geoffrey Chew was born in Washington DC in 1924. In 1940 he entered George Washington University as a chemistry major. Under the influence of George Gamow at GWU he switched to physics. After graduation in 1944 he was sent

to Los Alamos as an assistant to Edward Teller. Chew remained at Los Alamos until 1946. He became acquainted with Fermi and served as a teaching assistant in a nuclear physics course that Fermi taught at Los Alamos. He entered graduate school at Chicago and with Marvin Goldberger became a theoretical student under Fermi's direction. He received his Ph.D. in 1948. He served briefly on the faculty at Berkeley, but due to political discord he moved to the University of Illinois in 1950. He collaborated there with Francis Low, developing a model of pion-nucleon scattering which led the way to a more general S-matrix hadron dynamics. In 1957 he returned to Berkeley. With Stanley Mandelstam he developed a model for pion-pion scattering which ultimately led to the "boot-strap" idea that no hadron was fundamental—each being the composite of other hadrons being held together by still other hadrons. He was chairman of the Berkeley physics department from 1974 to 1978 and dean of Physical Sciences from 1986 until retirement in 1992. He has received many awards and prizes including the E. O. Lawrence Prize in 1969. In retirement he has continued research in quantum cosmology.

James W. Cronin

James W. Cronin was born in Chicago, Illinois, in 1931. He received a B.S. at Southern Methodist University in 1951. In 1951 he entered the graduate school at the University of Chicago and received a Ph.D. under the direction of Samuel K. Allison in 1955. After a three-year stint as a postdoctoral fellow at Brookhaven National Laboratory he joined the faculty of the physics department at Princeton University in 1958. In 1971 he moved to the University of Chicago. In 1997 he became professor emeritus and remains active in research at Chicago. In the years 1955–86 he worked in experimental particle physics, using most existing accelerators. Subjects of his research included pion-nucleus scattering, decay properties of hyperons and kaons, high transverse momentum phenomena, and a measurement of the lifetime of the neutral pion. In 1980 he and Val L. Fitch received the Nobel Prize for physics for their discovery of matter-antimatter asymmetry in neutral kaon decays. In 1987 he switched to cosmic ray physics. At present he is leading an international effort to study the nature and origin of cosmic rays with energies greater than 10^{20} eV. He has received a number of awards including the National Medal of Science in 1999.

George Farwell

George W. Farwell was born in Oakland, California, in 1920. He received a B.S. in physics from Harvard University in 1941. He joined the Manhattan District in 1941, working at Berkeley from 1941 to 1943, and at Los Alamos from 1943 to 1945. In 1944, with Emilio Segrè and Owen Chamberlain, he discovered the plutonium isotope 240. They found that the rate of spontaneous fission of plutonium 240 was so large that an implosion trigger was required for the plutonium bomb. In July 1945 he participated in the atomic bomb test at the Trinity site. From 1946 to 1948 he was a fellow of the Institute for Nuclear Studies at the University of Chicago, where he obtained his Ph.D. under the direction of Enrico Fermi. He joined the physics department at the University of Washington in 1946 and re-

mained there until his retirement in 1987. His research included studies of spontaneous fission, nuclear structure, time reversal invariance in nuclear reactions, radiocarbon in the environment, and paleoclimatology by means of radiocarbon dating of pollen in sediments using ultrasensitive mass spectrometry with accelerators.
He served in a number of capacities in academic administration including dean of
the Graduate School, director of the Division of Marine Resources, and vice president for research. George Farwell died in 2003.

Jerome I. Friedman

Jerome Friedman was born in Chicago in 1930. He received his A.B., M.S., and
Ph.D. degrees from the University of Chicago in 1950, 1953, and 1956, respectively. He began his thesis work under the direction of Fermi. After Fermi's death
his formal sponsor was John Marshall. After spending a year as research associate
in the Institute for Nuclear Studies at Chicago, he accepted a three-year appointment as a research associate at Stanford University. It was during his postdoctoral
year at Chicago that he with V. L. Telegdi showed that there was parity violation in
the pion-muon decay chain. In 1960 he joined the faculty at the Massachusetts
Institute of Technology and was promoted to professor in 1967. At MIT he has
served as director of the Laboratory for Nuclear Science and head of the Department of Physics. In 1991 he was appointed Institute Professor. He is an experimental particle physicist whose research has included studies of nucleon structure
and interactions of high-energy electrons, neutrinos, and hadrons with nucleons
and nuclei. In 1990 he was awarded the Nobel Prize in physics with Henry Ken-
dall and Richard Taylor for the experimental discovery of quarks in the nucleon.
He has received numerous other awards and citations. He has served on advisory
committees for the Department of Energy and laboratories throughout the world.
He was president of the American Physical Society in 1999–200.

Richard L. Garwin

Richard Garwin was born in Cleveland, Ohio, in 1928. He received a B.S. in phys-
ics from Case Institute of Technology, Cleveland, in 1947, and a Ph.D. in physics
from the University of Chicago in 1949. He is now Philip D. Reed Senior Fellow
for Science and Technology at the Council on Foreign Relations, New York, and
IBM Fellow Emeritus at the Thomas J. Watson Research Center, Yorktown Heights,
New York. After three years on the faculty of the University of Chicago, he joined
IBM Corporation in 1952 and was, until June 1993, IBM Fellow at the Thomas J.
Watson Research Center, Yorktown Heights, New York. He is adjunct research fel-
low in the Kennedy School of Government, Harvard University; and adjunct pro-
fessor of physics at Columbia University. In addition, he is a consultant to the U.S.
government on matters of military technology, arms control, etc. He has been di-
rector of the IBM Watson Laboratory, director of applied research at the IBM
Thomas J. Watson Research Center, and a member of the IBM Corporate Technical
Committee. He has also been professor of public policy in the Kennedy School of
Government, Harvard University. He has made contributions in the design of nuclear weapons; in instruments and electronics for research in nuclear and low-

temperature physics; in the establishment of the nonconservation of parity and the demonstration of some of its striking consequences; in computer elements and systems, including superconducting devices; in communication systems; in the behavior of solid helium; in the detection of gravitational radiation; and in military technology. He is the coauthor of many books, among them *Nuclear Weapons and World Politics* (1977), *Energy: The Next Twenty Years* (1979), and *Megawatts and Megatons: A Turning Point in the Nuclear Age?* (2001, with Georges Charpak). He was a member of the President's Science Advisory Committee, 1962–65 and 1969–72, and of the Defense Science Board, 1966–69. He has received many awards, among them the 1996 Enrico Fermi Award.

Murray Gell-Mann

Murray Gell-Mann was born in 1929 in New York, N.Y. He received his B.S. from Yale (1948) and Ph.D. from the Massachusetts Institute of Technology (1951). He is currently distinguished fellow at the Santa Fe Institute and author of the popular science book *The Quark and the Jaguar: Adventures in the Simple and the Complex.* He was on the faculty at the University of Chicago from 1952 to 1954, and professor at the California Institute of Technology from 1955 to 1993. In 1969, he received the Nobel Prize in physics for his work on the theory of elementary particles. His "eightfold way" theory brought order to the chaos created by the discovery of some one hundred particles in the atom's nucleus. Then he found that all of those particles, including the neutron and proton, are composed of fundamental building blocks that he named "quarks." The quarks are permanently confined by forces coming from the exchange of "gluons." He and others later constructed the quantum field theory of quarks and gluons, called "quantum chromodynamics," which seems to account for all the nuclear particles and their strong interactions. Although a theoretical physicist, his interests extend to many other subjects, including natural history, historical linguistics, archaeology, history, depth psychology, and creative thinking, all subjects connected with biological evolution, cultural evolution, and learning and thinking. His recent research at the Santa Fe Institute has focused on the study of complex adaptive systems, which brings all these topics together. He is also concerned about policy matters related to world environmental quality (including conservation of biological diversity), restraint in population growth, sustainable economic development, and stability of the world political system.

Maurice Glicksman

Maurice Glicksman was born in Toronto, Canada, in 1928. He studied engineering physics at Queen's University, entered the University of Chicago in 1949 to do graduate work, and received a Ph.D. in 1954 under the direction of Herbert L. Anderson. His graduate research studying the interaction of pions and protons was done in collaboration with Anderson and Enrico Fermi. Their discovery of excited nucleon states was a valuable marker in the development over the years of the standard model of strong, electromagnetic, and weak interactions. After a postdoctoral year at Chicago, he turned to a different research theme at the RCA Laboratories at Princeton, N.J., and Tokyo, Japan. He studied the properties of semiconductor and semiconductor alloy single crystals. He spent four years in Tokyo leading a basic-research laboratory operated by RCA. In 1969 he accepted an appointment as University Professor at Brown University. His research interests continued to include the electrical and optical properties of semiconductor crystalline material. In addition to teaching the normal undergraduate and graduate courses, he taught a seminar on the role of science and technology in the development of China and Japan. From 1974 through 1990 he held a number of senior academic administrative positions: dean of the Graduate School, dean of the faculty, and, from 1978 to 1990, provost. He retired in 1994. He has served voluntarily on boards and committees of a number of organizations, including Blue Cross of Rhode Island, the World Affairs Council of Rhode Island, and the international library cooperative OCLC, Inc. For sixteen years he served on the Board of Overseers of the Fermi National Laboratory. He was vice chairman or chairman of the board for eleven of those years.

Marvin Goldberger

Marvin L. Goldberger was born in Chicago in 1922. He grew up in Youngstown, Ohio. He received his B.S. in physics from the Carnegie Institute of Technology (now Carnegie Mellon University). In 1943 he was drafted into the army and assigned to the Manhattan Project as a member of the Army Special Engineering Project. He worked on the design of nuclear reactors in a group headed by Eugene Wigner. On discharge from the army in 1946 he entered the graduate school at the University of Chicago. He and Geoffrey Chew were Fermi's first theoretical graduate students at Chicago. He received his Ph.D. from Chicago in 1948. In 1950 he was appointed to the faculty at Chicago. While at Chicago he worked closely with Murray Gell-Mann on collision theory, statistical mechanics, and dispersion theory. In 1957 he joined the faculty of Princeton University. There he engaged in a variety of problems in the theory of elementary particles. Notable among these was the "Goldberger-Treiman relation," which showed an unexpected relation between the weak and strong interactions. In 1969 he became chairman of the physics department at Princeton University. In 1978 he became president of the California Institute of Technology. In 1987 he became Director of the Institute for Advanced Study. Since 1991 he has been Professor of physics at the University of California at San Diego, with a five-year stint as Dean of Natural Science. He has been actively involved as an advisor on national security affairs for the past fifty years. He was the founding chairman of the JASON group (advisors to the Department of Defense), a member of the President's Science Advisory Committee (1965–69), and the founding chairman of the National Academy of Sciences Committee on International Security and Arms Control in 1981.

Uri Haber-Schaim

Uri Haber-Schaim was born in Berlin in 1926. In 1949 he received an M.S. degree in physics from the Hebrew University in Jerusalem. He received his Ph.D. from the University of Chicago in 1951 under the direction of Enrico Fermi. His early professional career included work as a research physicist at the Weizmann Institute (1951–53), adviser to the cosmic ray group at the University of Bern, research associate at the University of Illinois, and member of the physics faculty at MIT (1956–61). In 1957, while at MIT, he became associated with the Physical Science Study Committee (PSSC), where he participated in the development of PSSC teaching materials. Since that time his attention has shifted to science education. He has been concerned with the problem of providing qualified teachers in the rapidly changing field of science education. To this end he has planned teacher training programs and conducted workshops in all parts of the United States and in foreign countries. He served as director of the Institute for Curriculum Development in Science and Mathematics at Boston University (1974–76). Under his direction the PSSC group developed *Introductory Physical Science* (*IPS*), a laboratory-oriented course for eighth and ninth grades (now in its seventh edition). He also directed revisions of *PSSC Physics* (now in its seventh edition). In 1993 he founded Science Curriculum Inc., which publishes the *IPS* and other science textbooks. In 1970 he received the Oersted Medal from the American Association of Physics Teachers.

Roger Hildebrand

At the time of the attack on Pearl Harbor, Roger Hildebrand was an undergraduate chemistry student at Berkeley. He was immediately put to work, first at Berkeley and then at Oak Ridge, on apparatus to separate uranium isotopes. At the end of the war he returned to Berkeley to complete a B.A. in chemistry in 1947 and then to earn a Ph.D. degree in physics in 1951. Soon after completing his formal education Hildebrand was sent by Ernest Lawrence to Chicago to do an experiment on uranium fission at the new Chicago cyclotron. This work led to a faculty appointment at Chicago in 1952. His first research as a member of the Chicago faculty was to study neutral pion production in neutron-proton collisions. He then constructed a liquid hydrogen bubble chamber, which he used for the first observation of the muon capture by protons.

In 1958 Hildebrand became associate director for high-energy physics at Argonne National Laboratory. In that capacity he was responsible for the construction of Argonne's 12.5-GeV Zero Gradient Synchrotron (directly managed by Albert Crewe). While at Argonne he continued teaching and research. With Peter Meyer he made the first detection of positrons in the primary cosmic radiation. After the ZGS was brought into operation, Hildebrand collaborated at Berkeley with Rae Stiening on experiments setting limits for a number of rare kaon decay processes. Hildebrand then switched to far-infrared astronomy. He developed polarimeters to map magnetic fields in interstellar clouds. Working with his students he discovered that in the far-infrared polarization depends strongly on wavelength.

At Chicago, Hildebrand has served as director of the Fermi Institute, dean of the college, and chairman of the Department of Astronomy and Astrophysics. His current research is an analysis of the characteristics of dust in the diffuse interstellar medium. He is currently the Samuel K. Allison Distinguished Service Professor Emeritus at Chicago.

Tsung Dao Lee

Tsung Dao Lee was born in Shanghai, China, in 1926. Following his sophomore year at Southwest Associate University in Kunming, China, he received a Chinese government fellowship for graduate study in the United States. He was accepted as a physics student in graduate school at the University of Chicago in 1946 and received his Ph.D. under the direction of Enrico Fermi in 1950. Following positions at the University of California, Berkeley, and the Institute for Advanced Study he was appointed to the faculty of Columbia University in 1953. With the exception of another period at the Institute for Advanced Study (1960–63) he has continued to be professor of physics at Columbia. In 1956 Lee, with C. N. Yang, made a thorough investigation of the evidence that nature is left-right symmetric and found there was no such evidence in any of the experiments on the weak interactions. This led to experiments that showed that left-right symmetry (parity) in the weak interactions was violated. Lee and Yang received the 1957 Nobel Prize for this work. Lee has made many contributions to particle physics, statistical mechanics, field theory, and astrophysics. He has also worked on the physics of relativistic heavy ion collisions, which are currently being investigated by the RHIC collider at Brookhaven National Laboratory. In 1980 he established the China-U.S. Physics Examination and Application program (CUSPEA). The program identified in China physics students with outstanding potential who would then apply for admission to participating U.S. graduate schools. The program was a great success, bringing more than nine hundred students to the United States for study. Since 1997 he has directed the RIKEN-BNL research center, which has developed high-performance computers for quantum chromodynamics calculations.

Darragh E. Nagle

Darragh E. Nagle was born in New York City in 1919. He received a B.S. in physics from the California Institute of Technology in 1940. He enrolled in the graduate program in physics at Columbia University and became associated with Enrico Fermi. He moved with Fermi to Chicago and participated in the construction of the first chain-reacting pile. He moved to Los Alamos in 1944 and participated in the development of the atomic bomb. Following the war he enrolled in the graduate program at MIT and received his Ph.D. in 1947. In 1949 he was appointed to the faculty at the University of Chicago. He participated in the pion-nucleon scattering experiments at the Chicago cyclotron. For these experiments he built a liquid hydrogen target. Following Donald Glaser's invention of the bubble chamber, he began to explore liquid hydrogen as a medium for a bubble chamber, and in early August 1953, with Roger Hildebrand, he demonstrated that superheated liquid hydrogen was radiation sensitive. In 1956 he moved to the Los Alamos Laboratory. There he first worked on experiments in plasma physics. He became the technical director of the 800-MeV Meson Factory at Los Alamos. More recently he participated in research aimed at identifying astrophysical point sources of gamma rays among the cosmic rays. He is a retired senior fellow at the Los Alamos National Laboratory.

Jay Orear

Jay Orear was born in Chicago, Illinois, in 1925. He received a Ph.B. in physics from the University of Chicago in 1944. This was followed by a two-year stint in the U.S. Navy, where he was an instructor in electronics. He returned to Chicago as a graduate student in 1946, receiving his Ph.D. under the direction of Fermi in 1953. With fellow students Art Rosenfeld and Bob Schluter, he wrote up the notes of Fermi's course on nuclear physics. Published as *Nuclear Physics* by the University of Chicago Press, it was a very successful book, which has been through many editions and still remains in print. Orear joined the physics faculty at Columbia University in 1954. In 1958 he joined the physics faculty at Cornell University, where he is now professor emeritus. His research at Columbia began with studies in nuclear emulsion of K meson decay. He contributed along with others to the tau-theta puzzle that led to the questioning of conservation of parity in the weak interactions. His research at Cornell was concerned with large-angle high-energy proton-proton and pion-proton scattering carried out at accelerators around the world. At the Fermilab Tevatron collider his group measured the total proton-antiproton cross section and elastic scattering at the highest energies. He has been active in the Federation of American Scientists, serving as chairman in 1967–68. He is the author of an unpublished report, "Statistics for Physicists," which is familiar to many physicists and is based on informal lectures by Fermi. He has taken an interest in the career and life of Enrico Fermi and is publishing a book on him.

Marshall Rosenbluth

Marshall Rosenbluth was born in Albany, N.Y., in 1927. He received a B.S. from Harvard University in 1945. He served in the U.S. Naval Reserve from 1944 to 1946. In 1945 he entered the graduate school at the University of Chicago and received his Ph.D. under the direction of Edward Teller in 1949. During the academic year 1949–50 he was an instructor in the physics department at Stanford University. While at Stanford he worked on the theory of electron-proton scattering, and the Rosenbluth formula for this process is well remembered. After leaving Stanford his scientific career was devoted to plasma physics. From 1950 to 1956 he was a staff member at the Los Alamos National Laboratory. From 1956 to 1967 he was senior research adviser to General Dynamics. In 1960 he also became professor of physics at the University of California at San Diego (UCSD). From 1967 to 1980 he held joint appointments at the Plasma Physics Laboratory at Princeton University and the Institute for Advanced Study. In 1980 he joined the faculty of the Institute for Fusion Studies at the University of Texas. In 1987 he returned to UCSD. He became professor emeritus in 1993. In 1992 he was appointed chief U.S. scientist for the International Thermonuclear Engineering Reactor (ITER). He was a consultant to the AEC, NASA, and the Institute for Defense Analysis. He received many honors and awards, including the Enrico Fermi Award of the Department of Energy, 1985, the National Medal of Science, 1998, and the Nicholson Medal for humanitarian service of the American Philosophical Society, 2000. Marshall Rosenbluth died in November 2003.

Arthur H. Rosenfeld

Arthur H. Rosenfeld was born in Birmingham, Alabama, in 1926. He received a B.S. in physics at the Virginia Polytechnic University in 1944. He received a Ph.D. from the University of Chicago in 1954 under the direction of Enrico Fermi. While at Chicago he with Jay Orear and Robert Schluter published the notes they had taken of Fermi's course on nuclear physics. The book, *Nuclear Physics,* was extremely popular and has gone through many revisions and is still in print. He became a research associate at the Berkeley Radiation Laboratory in 1955 and joined the University of California physics department in 1956. He became emeritus professor in 1994. His first research was with the Alvarez group studying elementary particle physics with bubble chambers. Rosenfeld was involved in the computer analysis of the bubble chamber data and the more general problem of data processing. He was instrumental in the foundation of the Particle Data Group at Berkeley, which continues to this day providing an archive for particle physics data. In 1974 he changed to a new scientific field concerned with the efficient use of energy. At the Lawrence Berkeley National Laboratory he formed the Center for Building Science. In the center, an electronic ballast was developed which led to compact fluorescent lamps. While at the center he developed the DOE-2 computer program for the energy analysis and design of buildings. He was appointed to the California Energy Commission in 2000. He has founded a number of organizations and institutes concerned with energy efficiency and its economic effects. He is the recipient of numerous awards and prizes: notable among these are the Szilard Award for Physics in the Public Interest in 1986 and the Carnot Award for Energy Efficiency from the Department of Energy in 1993.

Robert A. Schluter

Robert A. Schluter was born in Salt Lake City, Utah, in 1924. He served in the U.S. Army from 1943 to 1946. In the last two years of his army service he was assigned to the Manhattan Project as a member of the Army Special Engineering Project. His exposure to Fermi and others at Los Alamos led him to the University of Chicago, where he received his B.S. in 1947 and his Ph.D. under the direction of Fermi in 1954. While a student at Chicago he joined with colleagues Art Rosenfeld and Jay Orear in writing up the notes from Fermi's course on nuclear physics and publishing them in a book entitled *Nuclear Physics.* The book has been immensely popular and has gone through many editions and remains in print. He joined the MIT Laboratory for Nuclear Studies in 1955. There he worked with David Ritson on K mesons with nuclear emulsions. Also at MIT, he participated in the measurement of pion-proton total cross sections, which revealed some new resonances. He made an early attempt to measure the magnetic moment of the lambda hyperon. In 1961 he became an associate physicist at Argonne National Laboratory, and in 1972 he became professor of physics at Northwestern University. During this period his research was in elementary particle physics. This included experiments on K mesic X-rays in helium and production of charm. He later shifted his research to the hydrodynamics of fluid film motion in the presence of electric fields. He retired from Northwestern in 1992 and now resides in

Salt Lake City, where he is involved in efforts to improve mathematics instruction in the public schools.

Jack Steinberger

Jack Steinberger was born in Bad Kissingen, Germany, in 1921. He moved to the United States in 1934. He received a B.S. in chemistry from the University of Chicago in 1942. After his bachelor's degree he entered the U.S. Army and was assigned to the MIT Radiation Laboratory, where radar was being developed. After a period of active duty he entered the graduate school in physics at the University of Chicago. He received his Ph.D. degree under the direction of Fermi in 1948. He spent one year at the Institute for Advanced Study, where as a theorist he calculated the decay rate of the neutral pion, then a hypothetical particle. In 1949–50 he held a research position at Berkeley, where he discovered the neutral pion. In 1950 he joined the physics department at Columbia University. Using the Columbia cyclotron he studied the properties of the pions. In 1955 he began to build bubble chambers following their invention by Donald Glaser. With the bubble chambers he and his students did a comprehensive series of experiments which revealed the production and decay properties of the strange particles. In 1962 with Leon Lederman and Melvin Schwartz he showed that the neutrino associated with the electron is distinct from a neutrino associated with a muon. For this work he and his colleagues were awarded the Nobel Prize in physics in 1988. He then led a series of experiments on CP violation. In 1968 he moved to CERN, where he continued the CP experiments. Beginning in 1983 with colleagues from CERN, Dortmund, Heidelberg, and Saclay he carried out an extensive series of experiments on the determination of nucleon structure functions through neutrino interactions. His final experiments involved the construction of the Aleph detector at the Large Electron Positron Collider (LEP), which carried out precise experiments that verified the electroweak model. On his formal retirement from CERN in 1986, he became professor at the Scuola Normale, Pisa, Italy. Throughout his career he has been a strong supporter of nuclear disarmament.

Valentine L. Telegdi

Valentine L. Telegdi was born in Budapest, Hungary, in 1922. In 1946 he received an M.Sc. in chemical engineering in from the University of Lausanne. He received a Ph.D. in physics from Eidgenössische Technische Hochschule Zürich (ETH) in 1950. He became a U.S. citizen in 1957. He was appointed to the physics department of the University of Chicago in 1951. In 1972 he was named the Enrico Fermi Distinguished Professor. His research at Chicago was on many diverse subjects: photonuclear reactions, positronium, weak interactions (parity violation, beta-decay of the free neutron, muon decay and capture), muonium, g-2 of the muon, and kaon physics. This research was carried out at the Chicago cyclotron, the Argonne Zero Gradient Synchrotron (ZGS), CERN, and Fermilab. He also did theoretical work, for example, on the role of isospin in electromagnetic reactions with Murray Gell-Mann. In 1976 he became professor of physics at ETH, retiring in 1989. At ETH he continued research on weak interactions (second-class currents, muon neutrino helicity), atomic physics, and dimuon production by pions at the CERN Super Proton Synchrotron (SPS). He has served on many international committees, has held many distinguished lectureships, and is a member of the National Academy of Sciences and a foreign member of many science academies. He has received numerous awards including the Wolf Prize in physics in 1991 (shared with M. Goldhaber). Currently, he is a faculty associate at the California Institute of Technology and has a guest position at CERN.

Albert Wattenberg

Albert Wattenberg was born in New York City in 1917. He received a B.S. from the City College of New York in 1938. His first position in 1938 was as a spectroscopist for the Schenley Distillery Company. In 1939 he became a part-time graduate student at Columbia University and began taking courses from Fermi. In 1941 Fermi asked Wattenberg to join his group which was studying the feasibility of a chain reaction. Wattenberg moved with Fermi to Chicago, where he was a member of the group that produced the first controlled nuclear chain reaction in December 1942, where he was a witness of that historic event. He worked with Fermi's group throughout the war and in 1947 obtained his Ph.D. at the University of Chicago under the direction of Walter Zinn. In 1947, he began to work with Fermi and Leona Marshall on neutron experiments at the Argonne National laboratory. In 1949 he became a group leader at Argonne, where his research was in neutron physics and nuclear physics. From 1951 to 1958 he was a research physicist at the MIT Laboratory for Nuclear Science. In addition to research with the MIT electron synchrotron he contributed to the design of the Cambridge Electron Accelerator. With David Ritson, he studied K meson decays in nuclear emulsions exposed at Brookhaven and Berkeley. In 1958 he became professor of physics at the University of Illinois, Urbana. There he continued experiments with K mesons at Brookhaven and Argonne. At Fermilab, he was part of a team that produced the J/psi by photo production. He participated in charm physics experiments at the Stanford Linear Accelerator Center (SLAC). He was coeditor of Fermi's Collected Papers and is active in the history of physics, including being an editor of the History of Physics newsletter.

Frank Wilczek

Frank Wilczek is currently Herman Feshbach Professor of Physics at MIT. He was born in 1951 and grew up in Queens, Long Island. He graduated from the University of Chicago in 1970 with a B.S. in mathematics and got his Ph.D. in physics at Princeton in 1973. He has been a professor at Princeton University, the Institute for Theoretical Physics in Santa Barbara, Harvard, and the Institute for Advanced Study. His work covers a wide range of subjects in theoretical physics, mainly centered around quantum field theory and its applications. Especially notable is his discovery of asymptotic freedom, leading to the formulation and validation of quantum chromodynamics (QCD), the modern theory of strong and nuclear interactions. He has played a leading role in the development of these ideas and their extension to unified field theories. He has pioneered in the application of high-energy physics to early universe cosmology and the use of quantum field theory to describe many-body systems. He has won numerous prizes, both for his scientific work and for his writing. Since 1998, he has been a trustee of the University of Chicago.

Lincoln Wolfenstein

Lincoln Wolfenstein was born in Cleveland, Ohio, in 1923. He entered the University of Chicago in 1940 and received an S.M. degree in 1944. His master's thesis concerned the theory of extensive air showers. In 1944 he returned to Cleveland to work at NACA (the forerunner of NASA) on the theory of flow through jet engines. In 1946 he returned to the graduate school at Chicago. He did his Ph.D. thesis with Edward Teller on the theory of nuclear reactions with polarized protons. In 1948 he joined the faculty at the Carnegie Institute of Technology (now Carnegie-Mellon University). Most of his research has been on weak interactions. When CP violation was discovered in 1964 he proposed that it might be explained by a new superweak interaction. Experimentalists labored thirty-five years to discover evidence against this theory. The present description of CP violation is given by the 3×3 CKM matrix; in 1983 Wolfenstein presented a simple parameterization of this matrix, which has been extensively used. In 1978 Wolfenstein made the important observation that neutrino oscillations in a material medium could be quite different from those in a vacuum. Applied to neutrinos emerging from the sun by Stanislav Mikheyev and Alexei Smirnov, this became the MSW solution to the solar neutrino problem. He has spent much sabbatical leave at CERN and has been the recipient of a number of research professorships at U.S. institutions. In 1992 he received the Sakurai Prize from the American Physical Society. He became professor emeritus in 2000 and continues to work on CP violation and neutrino physics.

S. Courtenay Wright

Courtenay Wright was born in Vancouver, British Columbia, in 1923. He attended the University of British Columbia, receiving a B.A. in 1943. He served as a radar officer in the Royal Navy on the HMS Apollo, which was Eisenhower's command ship for the Normandy invasion. He entered graduate school at the University of California in 1946 and received his Ph.D. in 1949 under the direction of Emilio Segrè. He received a National Science Foundation postdoctoral fellowship in 1949 and moved to the Institute for Nuclear Studies at the University of Chicago. He was a research associate at Chicago until 1955, when he was appointed to the faculty of the Chicago physics department. He built a high-pressure diffusion cloud chamber, which was used for a number of investigations, including checking isospin predictions in pion production processes, and the electron spectrum in muon decay. He joined Roger Hildebrand in the construction and operation of a 15-inch bubble chamber, used for precision studies of pion production and decay. He used a polarized proton beam from the cyclotron to measure the spin-dependent coefficients of proton-proton scattering. With the advent of Fermilab in the early seventies, the Chicago cyclotron was dismantled, the magnet being moved to Fermilab to be used as a spectrometer for muon scattering experiments. Wright collaborated with Oxford and Harvard in studies of deep inelastic muon scattering. In the eighties he teamed with a group at Los Alamos to set a new limit on the forbidden muon decay to an electron and photon. From 1960 to 1968 he was a member of the JASON group, which gave scientific advice to the Department of Defense.

Chen Ning Yang

Chen Ning Yang was born in 1922 in Hefei, Anhui, China. He received his bachelor's and master's degrees during World War II from the National Southwest Associated University in Kunming, in southwestern China. At the end of 1945, he enrolled as a graduate student at the University of Chicago, having been attracted by the decision of Enrico Fermi to join that university. After receiving his Ph.D. under the formal direction of Edward Teller, he spent seventeen years, 1949–66, at the Institute for Advanced Study in Princeton. In 1966 he moved to the State University of New York at Stony Brook. There, he was appointed Einstein Professor of Physics and became director of the Institute for Theoretical Physics. He retired from Stony Brook in 1999 and is now associated with the Chinese University of Hong Kong and with Tsinghua University in Beijing. Yang's research in physics is mainly in particle physics, in condensed matter physics, and in statistical mechanics. In 1957 he was awarded the Nobel Prize in Physics with T. D. Lee, for the observation that the presumed left-right symmetry for the weak interactions had never been tested. This led to the experimental observation of a violation of left-right symmetry (parity violation) in the weak interactions. In his book *Selected Papers,* he said that the three physicists he most admired were Einstein, Dirac, and Fermi.

Gaurang Yodh

Gaurang Yodh was born in 1928 in Ahmedabad, India. He received a B.Sc. from the University of Bombay in 1948. He enrolled in the graduate school of the University of Chicago in 1948. He received his Ph.D. in 1955 under the direction of Herbert Anderson. His research involved pion physics at the Chicago cyclotron. He was an instructor of physics at Stanford from 1954 to 1956. In 1957 he returned to India to join the Tata Institute for Fundamental Research in Bombay. In 1958 he was appointed to the faculty of the Carnegie Institute of Technology. In 1961 he moved to the University of Maryland, where he remained until 1988. While at Maryland he was appointed visiting scientist at NASA. In 1988 he became professor of physics at the University of California at Irvine. For the two decades following his Ph.D., he worked in experimental particle physics. His research included measurements of pion production by electrons, pion production by pions, and the measurement of proton-antiproton total cross sections. He then shifted his interest to the study of energetic cosmic rays. This research included the measurement of cosmic ray composition in the 100-TeV range and the observation with cosmic rays that total cross sections were rising at energies above accelerator energies. Since 1985 he has been concerned with gamma-ray astronomy in the TeV range. Yodh is an accomplished performer on the Indian instrument the sitar.

Index